AF573832

HOW TO BUY AN ISLAND

other books by Donald McCormick

The Talkative Muse

Mr France

The Wicked City

The Hell-Fire Club

The Mystery of Lord Kitchener's Death

The Identity of Jack the Ripper

The Incredible Mr Kavanagh

The Wicked Village

Blood on the Sea

Temple of Love

Mask of Merlin

The Silent Killer

Pedlar of Death

The Red Barn Mystery

Murder by Witchcraft

Murder by Perfection

One Man's Wars

The Story of Charles Sweeny

HOW TO BUY AN ISLAND

by Donald McCormick

DAVID & CHARLES

NEWTON ABBOT

0 7153 5772 7

Set in eleven on thirteen point Baskerville
and printed in Great Britain
by W J Holman Limited Dawlish
for David & Charles (Holdings) Limited
South Devon House Newton Abbot Devon

CONTENTS

NOTE

Where US dollar conversions from pounds sterling are given, these have been based on an exchange rate of £1 = $2.40

LIST OF ILLUSTRATIONS

PLATES

ILLUSTRATIONS

MAPS

1 CAN YOU BE A CRUSOE?

Most of us possess some fondness for islands, and few of us have not at some time or other had an urge to own one, even though in most cases this wish was discarded in the nursery.

When I speak of islands, I mean the small unexpected island, the mere pinpoint on a map that most nearly fulfils our daydreams. I do not include large and over-populated islands, which are mostly tedious imitations of suburbia, or a combination of the Mad Hatter's tea party and a village fair. It is not easy to define the sort of island which, in this book, we shall try to find in fair measure, if not abundance, because to define is to limit, and the tastes of island-lovers vary widely. Then again, one must cater for the compromisers of this world—those who would prefer to have an island of their own, but, failing that, are willing to accept the alternative of life on somebody's else's island. But even they need a location that at least lends itself to the Crusoe touch.

Indeed, the actual existence of his dream island is not a prerequisite for the inveterate and fanatical island-seeker. If necessary, he will create it. Many have done so. R. M. Lockley, the ornithologist who took over Skokholm Island, a few miles off the Pembrokeshire coast, tells us in his book, *I Know an Island*, something of the spirit that moved him to create what he then lacked: 'The first island which I knew intimately,' he writes, 'was one made by hand. The ingredients were some alluvial soil and an oak tree about two hundred years old.' In the countryside of Monmouthshire he literally dug an island out of the soil that was piled up round an oak tree, and then, when he had moulded the plot to the desired size and shape, diverted a stream to make it an island in fact. This task, he adds, 'fed rather than appeased my hunger for islands. The idea remained in my mind that one day I might achieve my highest ambition and live on a real island of my own.'

Now this statement exemplifies the qualities needed by the island-seeker anywhere in the world. Escapism must be tempered by realism, the wilder dreams by the size of one's bank account; there must be sufficient flexibility of mind to turn away from the prospect of a palm-fringed coral reef thousands of miles distant to the lake at the bottom of the garden where, with sweat and effort, it may be more practical to create an island on a submerged mudbank.

This island complex is rooted firmly in our literature throughout the ages. So many of our best known books are about islands—*The Admirable Crichton, Robinson Crusoe, Treasure Island, Kidnapped, Dream Days, The Cautious Amorist,* even *The Tale of Squirrel Nutkin,* that inseparable companion of the nursery. All these have helped to mould the subconscious island philosophy that makes us escapists all in our innermost fancies and desires. In such a mood did Dorothy Osborne write to William Temple:

> Do you remember Herm and the little house there? Shall we go thither? that is next to being out of the world: there we might live like Baucis and Philemon and grow old together, and for our charity to some shipwrackt stranger, obtain the blessing of both dying at the same time.

It was the same spirit that caused Elizabeth Barrett Browning to sigh in verse

> My dream is of an island-place,
> Which distant seas keep lonely...

This train of thought and, indeed, my original *Islands for Sale* were inspired more than twenty years ago by a laconic paragraph in a newspaper. It stated prosaically that Lord Baldwin, son of the former British Prime Minister, had bought the island of Dead Man's Chest for £100.

Dead Man's Chest! How that name conjures up the romantic past of pirates, buried treasure, pieces of eight and multicoloured parrots. It has the authentic Stevensonian ring about it, and, in fact, was reputed to have given R.L.S. the inspiration for *Treasure Island*. What a superb address to give one's friends—J. Smith, Esq, Dead Man's Chest, c/o GPO, Antigua, West Indies. Of course, one would have to call one's home the 'Jolly Roger' and emboss the skull and crossbones on one's writing paper.

Lord Baldwin, the paragraph continued, had bought this island—the whole 25 acres of goat-inhabited scrub—to prevent its falling into the hands of a foreign buyer, and then had immediately presented it to the Virgin Islands' Legislative body. In 1950 a good many islands could still be bought for £100 and even less. It seemed at once to make island-owning less a romantic dream and more a practical proposition. It was not much more than a year's expenditure on cigarettes.

Was the island of one's dream so difficult to obtain? Was romance—a new adventure in living—one's own for the taking? I wanted to find out, and it was in the hope that others might also be anxious to know the answers to these questions that I wrote *Islands for Sale*. Having delved deeply into the problem of island home-hunting, I concluded that in those immediate postwar years of acute housing shortage, if one had enough adventurous spirit, it was no more of a problem than home-hunting in towns and cities.

Perhaps the frustrations of newlyweds, and people returning from the Forces after World War II and finding life dull, turned many of us into escapists. At this time, as a result of war and restrictions on movement, millions of people had not been able to travel abroad for the best part of ten years. Many of them yearned to do so. Those who had been in the Forces and had travelled on active service had been shown new horizons that had quickened their desire to see more. Yet global travel such as we now have in this jet age was then unknown. Above all, such wartime restrictions as rationing, exchange control and high taxation turned even the most humdrum citizen into a would-be escapist. As a result I received an enormous postbag for several months after the publication of the book, revealing that there were hundreds of people fascinated by the idea of buying, renting or acquiring an island or an island home.

The more recent island-seekers are, whether young or old, much less romantic than their predecessors. Indeed, romance to many has become, in the unlovely jargon of our times, an irrelevant word—partly because easier and quicker travel has debased the romantic currency of many islands that had hitherto only been visited by the occasional lotus-eater in quest of paradise. Islands such as Ibiza and Djerba which, in the early 1950s, conjured up for most people a prospect of what Vachel Lindsay called 'wizard islands, of august surprise', are now exploited by the package-holiday pedlars.

Thus, before tackling the subject again, I had to ask myself whether the basic premises of this whole question of island life were still valid. First of all, had the world, the property market and the pace of inflation changed so much that the idea of owning an island was even more of a pipedream for the average man than it used to be? And even if that was not so, were people now prepared to give up the benefits of the Welfare State, and the comfort of their automobile, television set, washing machine and spin-dryer to live on an island?

Let us take the economic aspects of the question first and try to answer the first query. In the past twenty years there has been a steady dwindling of population in most small islands all over the world. Therefore, if this trend continues, and if these islands become even more neglected, it is a fairly obvious economic fact that many of them are not going to maintain their value. So far, despite the dwindling populations, the prices of such islands have increased markedly and, possibly because of this, the demand has fallen. It seems logical to assume that this trend will ultimately be reversed, that is to say, there will be more uninhabited islands and more of them on the market at lower prices. When that point is reached, the islands will start to be peopled again.

The growth of the Welfare State and higher living standards have not dimmed the attractions of island life for the younger generation. If anything, bureaucracy, the rat race, over-population, regimentation, the spoliation of the countryside, the growing menace of pollution and ecological problems have spurred many of the young into experimenting with island communes and making a fetish of macrobiotic diets. So a new and revised look is required at the whole subject and this book is an attempt to bring the record up to date. We shall consider some 'luxury' prospects—we should not be escapists if we left them out—but equally we shall try to find a few 'bargain' islands for those with slender purses. Some will want to retire to their islands and bask idly in the sun from dawn to dusk. Others will find it necessary to look for somewhere that offers a chance of making a living.

In most cases it will be the state of your bank balance that will decide things for you. But do not be deterred from having an island merely because your job ties you to an office desk for most of the year. Small islands, whether you buy or rent them, need not be a costly investment: you can rent an island for as little as £10 a year, which over a long period would save you hundreds of pounds in holiday hotel bills. An island home can not only save you money but also give you the joys of escapism in which otherwise you would never be able to indulge. Apart from this an island can also be a hedge against inflation and provide you with sensational capital gains, especially if you improve it, despite my theory that sooner or later island prices will fall. Many islands have risen in value over the past forty years by more than 500 per cent.

This is, then, a kind of escapist's handbook, but, in defining it as such, it is important not to confuse escapism with 'ostrichism'. Escapism at its best is not turning your back on the world but rather stopping to savour the best it has to offer. Before reading further, you might try to decide what type of escapist you really are, and what you want an island home to give you, since the people who fail in this adventure are usually those with dreams they have never properly analysed or formalised.

If you are convinced that your present job is important enough to prevent any rash plunges, then concentrate on an island near home, or at least somewhere reasonably accessible. In these days of speedy air communication that is a simpler problem to solve than it was twenty years ago. If on the other hand you are the type who would like to retire to an island to spend the rest of your days, do not leave things to chance, hoping that, when the time comes for you to shut your desk and cast away your umbrella, you will find an island waiting for you somewhere or other. Finding the right island may well take years, for a large number of people have the same idea as you. So get down to the job of looking for an island before it is too late, even though you may be ten or fifteen years from retirement.

Beware of advertisements that say: 'Do you want to live in paradise? Here is an island, 30 acres, situated in the West Indies, which is yours for £4,500. Palm trees, tropical flowers and a perfect climate. Write Box——'. In recent years there has been quite a racket in the sale of white elephant islands and advertisements of this type have been fairly common. Subsequent inquiries have often revealed that the paradise so advertised is far from civilisation, having no fresh water supply, infested with sandflies and requiring as much as £15,000 ($36,000) to be spent on it before it can be considered habitable.

Sometimes there are genuine advertisements of islands for sale, but, generally speaking, all advertisements call for very careful investigation. When inspecting a house you propose to buy, you look out for dry rot, lack of damp courses and inadequate sewage disposal. With an island you look first and foremost for fresh water supplies, then for pests, freedom from flooding, adequate landing-sites, etc. Occasionally there are bargains in islands as with houses. Shortly before World War II an island in the Channel Isles was offered for sale in *The Times* for a mere £5—a bargain if ever there was one, for it has since changed hands at more than £25,000 ($60,000).

A word to would-be beachcombers. The days of Gauguin and other romantics are over. There are not many islands today where beachcombing can be carried on either in its literal and original sense or the modern interpretation of 'living off the country'. To begin with, the cost of living on most islands has risen and, while it may be much cheaper than the mainland (though quite often not), it is not cheap enough for beachcombing. There was a time, of course, not much more than 100 years ago, when the inhabitants of the Isles of Scilly pursued their own interpretation of beachcombing, which was to live off the many wrecked ships cast upon their shores. But everywhere today beachcombers of all types are being discouraged by regulations designed to make their kind of life almost impossible. This is especially true of such places at Tahiti, where the authorities want all sorts of guarantees before they permit immigrants to settle. Indeed, they insist that even the casual visitor must have a return ticket and adequate funds for the period of his stay. Similarly, it must be stressed that you cannot find islands today where your lunch falls off a tree into your lap. You must at least fish for it, or do some honest digging with a spade.

I have gone into this question of beachcombing thoroughly because I find that island-hunters are often incurable optimists who believe that if they paint a few pictures, or play a guitar whenever they are thirsty or hungry, some benevolent islander will befriend them. If Gauguin found it a dismal failure, there is no reason to suppose that amateurs would succeed. Conditions for the sublime if impoverished optimist are likely to be worse in the near future. From Katmandhu to Ibiza the powers-that-be are tightening up regulations. In the past year the Spanish authorities in the Balearics and the Greek police in the islands of their own archipelago have made a concerted drive against hippies, dropouts and those more or less begging for a living.

Yet one must admit that all this does not deter the beach-comber-hippie fraternity. Indeed, while some of their experiences may be a warning of what to avoid, they are equally an indication of what their seldom publicised courage, initiative and refusal to admit defeat can achieve. Three years ago the Diggers' Action Group was given the island of Dorinish, 6 miles off the coast of Mayo in west Ireland, by pop singer John Lennon. Thirty people landed there and declared themselves the possessors of total freedom. Though their romantic dreams may have been somewhat dinted, at the time of writing they remain there, reduced only to twenty-five.

The *Sunday Times* reported on 5 September 1971 that Dorinish was

> . . . 25 acres when the tide is out, but considerably less when it is in. . . The inhabitants call themselves the Tribe of the Sun. . . Nearby three men are tending to the vegetable fields and outside the main mess tent, which is surrounded by turfs to protect it from the wind, a man is playing the flute. . . After surviving one cruel winter the men and women who came with their hands and virtually no money are facing another winter when it is possible they will either leave or perish. . . The water wells are running dry and they need plenty of water; they have got no money and need it to get materials from the mainland; some of the locals believe they are growing hashish and the police visited them once. They now need £2 per week per head to keep going and last year they needed £1.50; inflation, it seems, has even got to Dorinish.

For those who take the quest for islands in more business-like fashion and want to know how much it costs to buy one, my estimates of twenty years ago—£20 to £20,000—have increased to £100 to £100,000. You could not expect to get anything glamorous for £100 and you could still pay much more than £100,000—half a million would not be the absolute limit ($240,000). Yet there is still a fairly wide range of islands to suit an equally wide range of incomes. What you have to consider is not so much the cost of the island as how much you will need to spend to make it habitable.

Before you settle on your island, you must be sure you can cope with all its day-to-day problems, which you deal with on the mainland by telephoning plumbers, electricians, Council offices and builders. Furthermore, if you do not possess independent means, a pension or some other form of unearned income, you will require to have such technical, professional or other skilled knowledge that will enable you to make a reasonable living in your chosen island home. There are not many professions that can be practised as easily, say, on an outpost in the Pacific as in Europe or the USA. An established author might succeed; it is probably the best answer to his problems to find a tax-free island in which to write without the interruptions and discords of city life. But there is always the risk that being out of touch with the people to whom he sells his work will ultimately tend to reduce his income. It is far wiser to treat such a livelihood as writing, painting, or any other artistic calling as a useful sideline to supplement a private income.

Yet enterprise pays even on an island. You need not be dismayed by the problem of earning a living on an outpost. In the Great Barrier Reef off the Australian coast a young airman bought up a few planes and established a Barrier Reef Airways Service. In another of these islands there is a flying doctor whose practice is larger than the whole of the British Isles. He also includes dentistry as a profitable sideline. There is no reason why many islands cannot be modestly exploited for tourist purposes, an idea which may be anathema to the romantic escapist but may make all the difference to the comfort and amenities of life. After all, if you love fishing, for example, why not encourage fishermen to come and pay to fish with you?

Farming and horticulture should be regarded as possible sidelines, though only for pin-money, for the day of the small-scale farmer is almost ended. Islanders will always find it difficult to market their produce, but a few pigs, a goat or two, chickens and some crops to cover all seasons can at least help to reduce household budgets and even enable you to cope with occasional visitors more hospitably than is possible on the mainland.

The part played in empire building by some escapists turned pioneers is in itself a fascinating story, but not normally feasible today. Yet it is surprising how many of these early islanders made fortunes out of uninhabited wastes. The Clunies-Ross family made theirs from the phosphate deposits of Christmas Island in the Indian Ocean (see p150), and there are possibilities for entrepreneurs in island-owning today; but these are the exceptions and one must remember that most islander home-hunters are not seeking a fortune.

It takes much longer to find the right island today than twenty years ago. It is for this reason that I have given as a practical price range £100 to £100,000. You can look outside this range but, while you may strike lucky, you could also hit a great deal of trouble. My figures will be disputed by many. For example, it is possible to buy a 1 acre island in the Canadian waters for as little as $60. On the other hand a leading estate agent who deals in islands insists that 'islands are getting scarcer' and that his list ranges from '£5,000 to £2,750,000'. Anything offered at more than half a million is either grossly over-valued, or contains deposits, minerals or other commodities that bring it into the realms of big business.

If you are looking for a cheap island, a good rule is to find it yourself rather than relying on an agent or answering advertisements. If you are prepared to pay upwards of £2,500, however, then go to an agent and see what he produces. Some advice on agents will be given for each area we explore. Outside the sterling area you will need first of all the permission of the Bank of England, which allows each family to own only one foreign property. Other countries' regulations vary somewhat, but in most areas of the world there are regulations to comply with if you wish to purchase property outside the jurisdiction of that country. You will, of course, need to buy the necessary local currency to pay for your island. This can be very costly, since, under the Exchange Control Act of 1947, gold and foreign currency reserves are not available for the purchase of property overseas. You must buy your money from a special currency market—the investment currency market. The premium can run up to 20-30 per cent more than the cost in the official exchange market. It is wise to have a solicitor acting for you right from the start. Cases have occurred where the offer has been a hoax and the property not really for sale.

The islands that are becoming scarcer on the market are the better known ones and the most desirable of properties. Provided you are not too fussy, and provided you are a dedicated island-lover and not merely one who wants an island plus all the luxuries of civilisation, you should concentrate your search on the many islands that are steadily being vacated or whose already small populations are dwindling.

Now let us set off exploring.

2 THE BRITISH ISLES

SOUTH AND WEST BRITAIN

AROUND the British Isles there are more than 5,000 islands of varying sizes, a revelation that surprised me as much as it will probably surprise you. The vast majority of them, it is true, are little more than rocks or barely visible patches of sand, and may not technically qualify for the title of island; but if you are keen enough, and especially if your purse is limited, a small rock for £20 or £50 will probably satisfy your yearnings, providing you do not wish to live there permanently.

Let us first consider south-west Britain, comprising the west and south coasts of Wales, the Bristol Channel and the Isles of Scilly. Here are islands of all shapes and sizes to suit all pockets. There are very few truly sheltered islands round the British coasts. To live in any of them one needs to love the sea in all its most tempestuous moods, and the temperament to stand days of raging gales. But for warmth and abundance of sun the islands of south-west Britain offer far more attractions than most of the others.

R. M. Lockley, mentioned in Chapter 1, achieved his ambition to have an island of his own when in 1927 he took Skokholm (see p33), off the south-west corner of Wales, on a 21-years lease at £2 a year. The previous history of Skokholm is interesting. When the Earl of Pembroke died in 1524, the 'valor' of his estate included £2 15s for the pasturage of the three islands of Skokholm, Skomer and Midland. Skokholm was frequently seized by the king during its early history as a

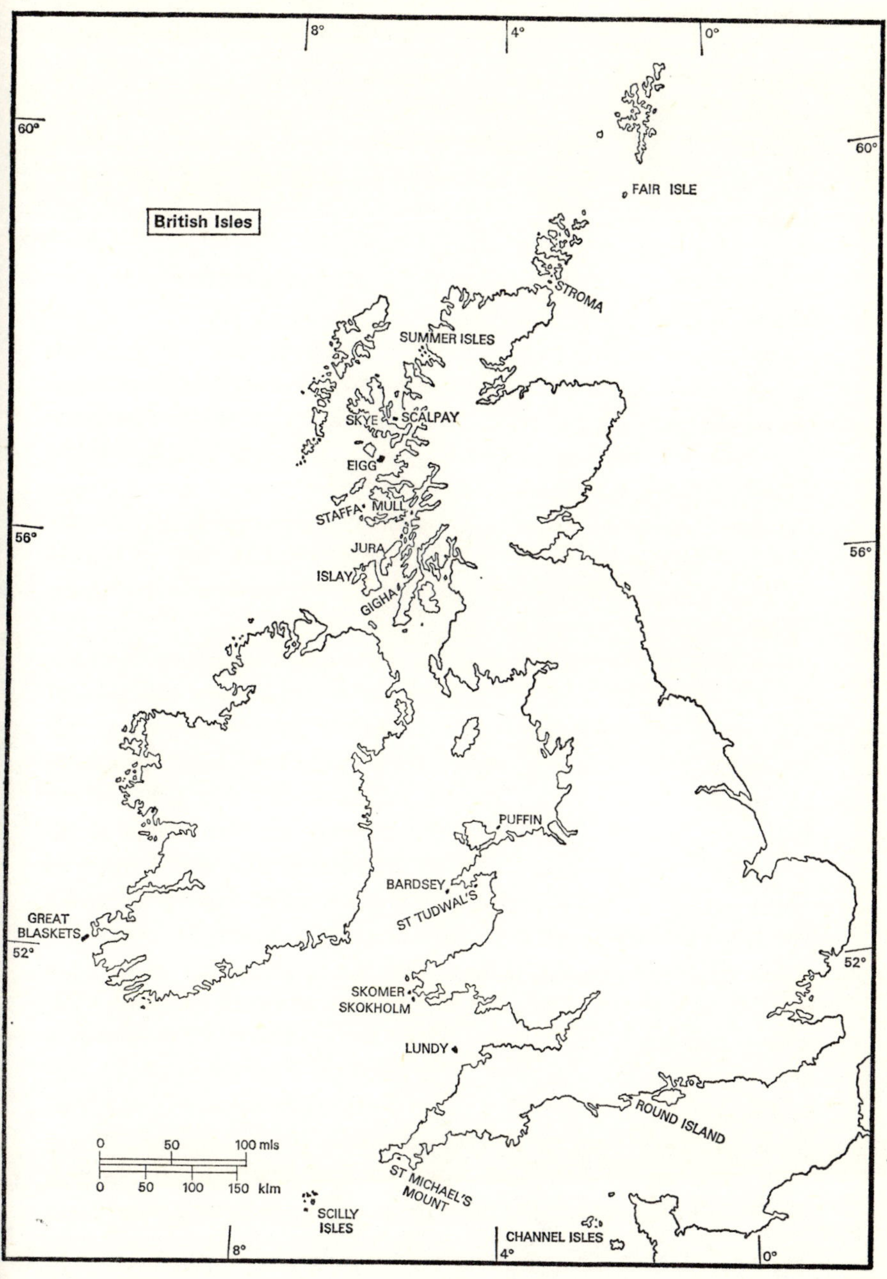
British Isles
8°
4°
0°
60°
60°
FAIR ISLE
STROMA
SUMMER ISLES
SKYE
SCALPAY
EIGG
STAFFA
MULL
56°
56°
JURA
ISLAY
GIGHA
PUFFIN
BARDSEY
ST TUDWAL'S
GREAT
BLASKETS
52°
52°
SKOMER
SKOKHOLM
LUNDY
ROUND ISLAND
0
50
100 mls
0
50
100
150 klm
ST MICHAEL'S
MOUNT
SCILLY
ISLES
CHANNEL ISLES
8°
4°
0°

punishment for the treason of one or other of the Earls of Pembroke, and documents suggest that an excellent income was made out of the rabbits that swarmed over the island. In those days there was a community living permanently on the island and they seem to have eked out a bare living. Bad weather was their chief problem and it is recorded that they lit fires on Bread Rock when flour was wanted and on Spy Rock when a doctor was urgently needed.

Skokholm is 240 acres in area. Mr Lockley found a house, which was in need of repairs, and the early months of his sojourn on the island were spent in stopping leaks in the roof of the house, clearing the ground and making a slipway for boats. But he had much to look forward to during his early months of pioneering, for, as he tells us in his book *I Know an Island*: 'Before I left Monmouthshire a courageous young woman there had promised with the greatest eagerness to tackle this island experiment with me. I was to have a house at least roofed for her by the day in July of next summer which had been fixed for our wedding. But I was instructed to leave as much of the pioneering as possible for her to share in after that date'.

Perhaps fortune favours those who think on these lines and act on their hunches. At any rate, fortune favoured Mr Lockley to the extent of causing a wooden topsail schooner, *Alice Williams*, to ram herself in a heavy fog on the east side of Skokholm. For a mere £5 he bought the ship intact from the underwriters. The figurehead of the wreck was fixed on Skokholm as an anchor mark at the entrance to South Haven. The ship's galley and wheelhouse became shelters for tools and hens, and the freshwater tank proved an absolute godsend. The wrecking of that ship—the crew, luckily, were saved—was an untold blessing for Mr Lockley.

There are innumerable islands like Skokholm to choose from off the west and south-west coasts of Wales and off the south-west tip of England. Some can still be bought, others leased. The real bargains are rarely advertised and usually are the least known islands, so go out and look for them, but methodically. First decide upon the area in which you are going to make your search, buy up all the large-scale ordnance maps covering the area, then consult the Admiralty Pilot books, which give detailed descriptions of islands off all coasts throughout the world and index them alphabetically. It will also help to consult Admiralty charts of the area. By this means you will probably discover a number of tiny islands that do not normally appear even on large-scale maps. It is often much easier to buy an uninhabited island than one that has already been made habitable. It is also possible that, if you find such an island, its owner will never have thought of selling or renting it.

Just as the village pub is as good a place as any to pick up some gossip on derelict cottages that can be bought for the proverbial song, so a fishermen's waterside tavern is the ideal venue for seeking out information on local islands. Put on an old pair of seaboots, a sweater and a pair of baggy trousers, avoid any suggestion of possessing much money, and then casually inquire what the prospects are of renting 'that old rock in the bay'. You might strike lucky.

Many of the islands in south-west Britain, as elsewhere in the British Isles, have special privileges to offer the true escapist. Some constitute civil parishes of their own—Skokholm is an example of this—which means that the owner is automatically chairman of the parish council and executive in one, bringing him into touch with Whitehall in an almost mandarin-like capacity. Some of the larger islands enjoy even more privileges and correspondingly more freedom. If you merely wish to live on an inhabited island where there is more freedom than on the mainland, there is a choice. Lundy in the Bristol Channel is as good an example as any. Here there are no rates, taxes, tithes or Inland Revenue duties, and wines and spirits can enter without excise charges. The Marisco Tavern there is licensed for the whole of twenty-four hours.

Lundy, which lies 25 miles west of Ilfracombe, was bought

for £16,000 in 1925 by Mr Martin Coles Harman. When it was presented to the National Trust in 1969, thanks to the generosity and public spirit of Mr Jack Hayward, a Bahamas property developer, he paid £150,000 for it, which is a further example of the rising cost of islands. The present population of the island is small for its size, twelve in all, but for anyone wanting to enjoy peace, solitude and as much freedom as you are likely to get anywhere in this world, Lundy offers unlimited possibilities. Those interested in settling there (very occasionally jobs are advertised on the island) should make inquiries of the Secretary to the Landmark Trust, Shottesbury Park, White Waltham, Berkshire. The main disadvantage is access to the island. Boats sail to Lundy from Ilfracombe and the crossing takes about 3hr, but sailings depend on the state of the weather. It is possible to be marooned on the island for days if there are gales.

One of the chief objections of intending Crusoes to islands round the British coasts is that they are nearly all cold, bleak and isolated. But they are wrong: 40 miles south-west of Land's End lies a group of islands which, though isolated, have mild winters and hot summers, and where the months of February and March display a carpet of brilliant flowers.

Take a look at the map on page 27. There are 145 of these islands in the Scillies, some little more than rocks, but they have the advantage of being clustered together like chickens round their mother, within easy reach of the main islands of St Mary's and Tresco. Access to them is simple enough—by regular boat service from Penzance, or by helicopter from the heliport in the same town—a mere 20 minutes' hop across the sea. This is the one part of the British Isles that unfailingly gets an early spring, a spring that rides in on the Gulf Stream. February is always a delightful month. You can leave an icebound London, catch the Cornish Riviera express, arrive at Penzance and within half an hour find yourself bathed in sunshine.

The Scillies are the property of the Duchy of Cornwall,

which occasionally leases individual islands and whose offices are in Buckingham Gate, London. For local government purposes the islands have a council of their own. But, alas, the islands are no longer exempt from paying income tax! That concession was withdrawn from them by R. A. Butler when he was Chancellor of the Exchequer in 1954. Rents have certainly gone up here as elsewhere, but by mainland standards they are still incredibly cheap. The inhabitants, who greatly resented the removal of the income tax concessions, point out, however, that the cost of living is higher than on the mainland. Water pumped from a well can cost 75p a 1,000 gallons. Freight charges increase the cost of goods, making bread and water substantially dearer than they are in Cornwall.

The inhabited islands—St Mary's, Tresco, St Martin's, St Agnes and Bryher—all possess the basic amenities of life. Each has its own post office, general stores and school. St Mary's has a hospital, two dozen shops, banks, a golf course, cinema and various clubs and societies. Tresco has a hotel and an inn. There are cottages to let on all these islands.

Some of the uninhabited islands are scheduled as bird sanctuaries and therefore cannot be invaded by would-be homemakers. But, as I have said, there are 145 islands in the Scillies, and many of those now uninhabited at one time supported fair-sized communities—so they are capable of development, even though the population of the smaller inhabited islands has been gradually declining for some years.

The prosperity of the islands has for many years depended on the flower industry. Scilly has always scored heavily with its early season blooms for Covent Garden, more than 1,000 tons of flowers being shipped annually to the mainland. Experiments in the growing of new varieties of flowers are continually being made. But, a word of warning to those who may be entranced by the possibility of flower-growing on the islands: it is generally conceded that when the United Kingdom enters the Common Market, flower-growers here will tend to suffer most.

You can go from island to island by rowing boat, or sometimes walk at low tide. *En route* you have the added satisfaction of being able to buy lobsters at bargain prices from the fishermen, or a large dish of crabs fresh from the sea. But the greatest delight of the islands is the flower-gardens. In springtime they create a veritable carnival of colour, and the patchwork of oddly shaped bulb fields makes a brilliant pattern of yellow, gold and white. The flowers extend beyond the hedges of the fields on to the seashore. Around March the daffodils give place to red camellias.

From Land's End to Poole Harbour there are many tiny islands along the coast, and some of them from time to time come on the market. There are five islands in Poole Harbour alone, the best known being Brownsea, which now belongs to the National Trust. The others are privately owned. Round Island, comprising 9½ acres, has six houses with electricity and water laid on. It was offered a few years ago at £50,000, this price including a pier, boathouse, cabin launch and a 30ft motorboat.

A much cheaper island, which came on the market in the mid 1960s, was South Brent Island, situated in the River Avon, 20 miles from Torquay in Devon. This 2½ acre island is connected to the mainland by a weir and a bridge, and its asking price then was £2,500, with salmon and trout fishing rights thrown in.

Smaller, somewhat bleaker islands are also to be found off the coast of Cornwall and in the Bristol Channel. Such islands do not easily come on the estate agents' books and inquiries about their availability must be made on the spot.

Farming on the islands in south-west Britain and north-west Wales generally offers only a supplementary income, though in some rare islands farming is more profitable than on the mainland. Bardsey Island, off the south-west corner of Caernarvonshire and in sight of Snowdon, was for years a case in point. It has excellent land, though it is at the mercy of incessant storms in winter, and owes its relative prosperity to

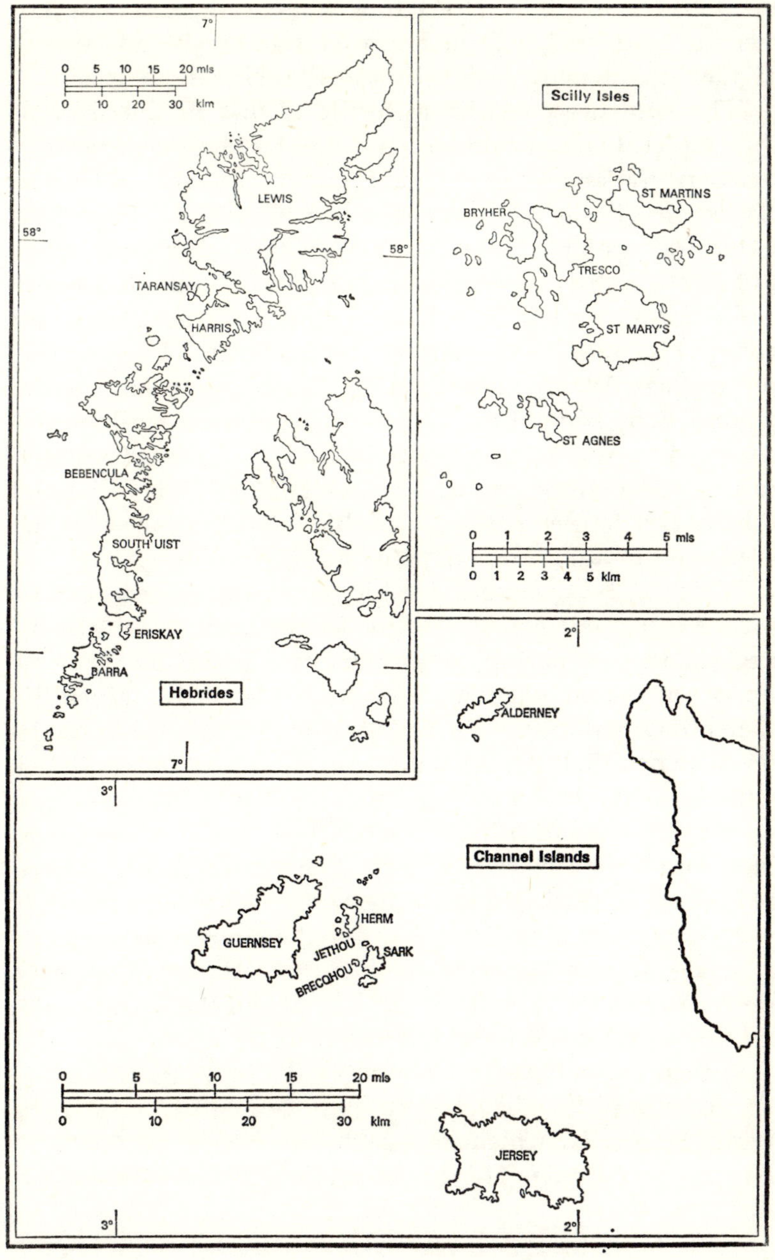
Hebrides
LEWIS
TARANSAY
HARRIS
BEBENCULA
SOUTH UIST
ERISKAY
BARRA
0 5 10 15 20 mls
0 10 20 30 klm
7°
58°
Scilly Isles
BRYHER
TRESCO
ST MARTINS
ST MARY'S
ST AGNES
0 1 2 3 4 5 mls
0 1 2 3 4 5 klm
Channel Islands
ALDERNEY
GUERNSEY
HERM
JETHOU
SARK
BRECQHOU
JERSEY
0 5 10 15 20 mls
0 10 20 30 klm
3°
2°

the hill that dominates the green pasture land at its foot and breaks the strength of the Atlantic gales on the west. To protect the land further from storm damage the island farmers divided it with many hedges into small sheltered fields.

'The land here is twice as fertile as that in Caernarvonshire,' a local farmer told me. 'It is also far cheaper. I came to this barren island simply to make money.' The rent of land in Bardsey has in the past been as low an 10s an acre. The whole island was up for sale for £50,000-£60,000 a few years ago—440 acres, twelve houses and cottages, four farms, a boathouse and slipway, a small chapel and a ruined thirteenth-century abbey. The buildings are strong, commodious and better than the average Welsh farm dwelling. Each house has on the ground floor from three to five living rooms, scullery and dairy, and upstairs three to five bedrooms. A motorboat used to take the island's produce to the mainland. There is good fishing and lobster hauls are prolific, and, like most of the islands in this area, it is a paradise for the birdlover.

What can be a pleasantly habitable island for years can be turned into a setting for a Gothic horror story by subsequent neglect. In considering any uninhabited island that may be for sale or to rent, one must study at first hand the animal life thereon. One thinks of Puffin Island only ½ mile from Anglesey as an ornithologist's paradise, but it is in fact inhabited by something like half a million rats—an awful warning of how an insidious rat invasion, if unchecked, can turn an island idyll into a nightmare existence. The rats originally arrived off a wreck in 1816, and today the people of Anglesey fear they are too numerous and too close for their own comfort. The rats have, in fact, destroyed the puffin population. They have slowly eaten their way across the island, and the whole area is riddled with rat holes and burrows.

Rather more happily situated is St Tudwal's Island in Cardigan Bay, less than 1 mile from the southern tip of the Lleyn Peninsula, which has magnificent views of Cardigan Bay and Snowdonia. This island of about 20 acres has a stone-built lighthouse, which was considerably improved and turned into a dwelling by Giles St Aubyn, son of Lord St Levan, who owns St Michael's Mount off the Cornish coast. Island-owning seems to run in the family. There are no water worries here, as an unfailing supply of copious Welsh rain is piped from the roof into three large underground tanks. Electricity is provided by a generator and cooking and water-heating are done by calor gas. The island also has a granite quay maintained by Trinity House. In the late 1960s the island was offered on an 89-years lease for £9,500.

THE CHANNEL ISLES

For those who wish to compromise on island home-making by selecting an isle within reasonable distance of the British coast yet possessing a life that is almost Continental, then the Channel Isles are the obvious choice. This geographical anomaly of a Continental semi-feudal entity within the British Isles offers an escape to a new life in pleasant and not too expensive surroundings. The fact that the cost of living in Jersey is estimated to be about 30 per cent higher than London's confirms the argument that statistics can prove anything. Food costs more than in Britain and so do some other things, but income tax is very low and there is no purchase tax. So, for reasonably well-off people, the cost of living is probably cheaper than in Britain. In the other Channel islands it is lower still. You can enjoy French cooking and wines far more cheaply than on the mainland and buy cigarettes, tobacco and spirits without feeling the crunch of punitive taxation. You can also enjoy the freedom from the British mainland's irksome licensing restrictions. Jersey has a remarkable range of entertainment and recreation, and the smaller islands have the remoteness of the Middle Ages. When you feel in need of a fuller life, the French coast is only 20 miles away and you can easily get to St Malo or Cherbourg.

Novembers can be sunny in the Channel Isles, and the

climate, despite some fierce storms that drive up the English Channel, is mild. In Guernsey semi-tropical plants grow all the year round and bathing is good until well into autumn.

Immediately after World War II there was a rush of capital and people from Britain to Jersey and, it must be admitted, both the capital and the people were in some cases rather suspect. In consequence, stringent regulations were introduced —and these should be borne in mind by anyone thinking of settling in Jersey or Guernsey—to control the acquisition of property and immigration. If you live in a hotel or boarding house, you can stay as long as you like and become a permanent resident. But to buy or rent a house, you must not only get permission from the authorities, but so must the person with whom you propose to make the deal.

This last regulation does not yet apply to Alderney and the other islands, though Alderney is now preparing legislation to control immigration and housing. A house on Sark occasionally comes on the market, the annual lease for a site of a fifth of an acre ranging from £50 to £100. But in Jersey and Guernsey property is very much higher in price. No house can be bought here if an islander born in Jersey is able and wants to buy it, which means that few of the cheaper houses are available to newcomers. Outsiders can buy from about £10,000 upwards, though one has to approach £25,000 before one can get much of a choice. But if you are able to buy a house in either of the two main islands the rates will be far lower than in the UK: many country houses with four bedrooms and two or three living rooms pay not much more than £12 a year in rates.

The islands' Health Service is very similar to that of Britain prior to 1948 and it is generally advisable to take advantage of private health and medical insurance schemes. The mild climate reduces heating costs considerably. Spirits, wine and tobacco are all remarkably cheap, whisky costing about £1.30 a bottle, gin £1.10 (there is even a local gin at 80p) and English cigarettes are about half the price paid in Britain. The standard rate for income tax in both Jersey and Guernsey is 20p in the pound.

The main attraction of the islands is the variety of scenery packed into so small a space. This applies particularly to Jersey and Guernsey. You have a series of cosy little bays, long stretches of perfect sandy beaches, woods running down to the seashore, orchards, furze and heather-covered hills. The lanes of Jersey meander past farmhouses with ivy-covered Norman arches and many of the houses have the initials of the owners engraved on the stone arches, with a pair of hearts entwined in the centre of the engravings. In spring the islands are carpeted with an astonishing variety of flowers, ranging from bluebells and large creamy-headed daisies to the deep crimson of red campion.

Both Jersey and Guernsey provide as varied attractions as the average resort in the British Isles, including organised deep-sea fishing, golf, tennis, bowls, yachting, riding, theatres, cinemas and concerts. Alderney, 3sq miles, with a population of about 1,600, is an ideal centre for fishing, and besides the main islands there are Sark, Herm, Brecqhou and many others.

Herm (see p33), a tiny isle about 8 minutes by speedboat from Guernsey, has a beach formed by a solid deposit of minute shells that have drifted in with the Gulf Stream. The insularity of Herm appealed to Sir Compton Mackenzie, who wrote:

> Desirable islands are rare. I had long ago marked down Herm as a desirable island, the possessor of which might one day compete with Tresco, and Tresco has the finest garden in Europe. One of the characteristics of island-lovers, a noble race belonging to the youth of the world, is an ability to make up the mind quickly and to change it as rapidly . . . the island was acquired.

The change of mind to which Sir Compton was referring was that he had postponed a world trip in order to buy the island. His reference to Tresco, the most beautiful of all the Scilly Isles, concerns a wonderful garden of many tropical and subtropical plants and flowers created by Augustus Smith in the nineteenth century.

Herm, with its fairy-story miniature harbour, owes its pines and cypresses to the Trappist monks who used to live there. In summer the island has a dazzling display of primroses, bluebells and foxgloves and Compton Mackenzie himself introduced many new flowers to the garden he created round his cottage. Herm also offers birdwatchers the chance to study curlews, linnets, puffins and kingfishers.

Herm's various owners have all enriched its history. Prince Blucher once leased the island and brought kangaroos with him, but they are now extinct. When Compton Mackenzie left there, Lord Perry, of the Ford Motor Company, became the tenant; then in 1946 the Treasury announced that the island was to be let for £900 a year—a high rental at first impression, but it included one large residence, twelve houses and a modern farm. Some months later Herm was sold by the Treasury to the Government of Guernsey for £15,000 after the leasehold at £9,000 had been on the market without attracting a firm bidder. The only inhabitants then were a caretaker, a 'constable' and his wife and several hundred rabbits, though once the island had a population of more than fifty people.

In seeking islands one can learn a great deal from others and nowhere perhaps does this maxim apply better than on Herm. By 1949 its rich natural vegetation, intermingled with eucalyptus, palms, Japanese cactus, silver poplar and gum trees, had run wild. Buildings such as the chapel of one mysterious missionary known as St Tugual were falling apart. Then a former British Army officer, Major A. G. Woods, and his wife became the tenants and set about restoring order. The island was so overrun with weeds and plant life that it took them three weeks to cut a path to the house in which they were to live, and another 5 years to cut paths to various points on the island.

Major Woods struck a bargain: he arranged over a three-

Page 33 (above) *Skokholm off SW Pembrokeshire;* (below) *Herm Island, with Jethou centre and Guernsey in the background*

Gigha, paradise of flowers off Argyllshire

year period to pay not £900 a year rent but £10, and after three years to put his accounts before the Guernsey authorities. After that initial three-year try-out he made a new arrangement: Guernsey agreed to continue helping to restore and develop its property on the island and to make a 'subsidy' of £3,000 a year to Major Woods. In fact Guernsey made a good bargain, for the 'subsidy' was covered by a levy on people who paid 25p a head for a trip to Herm by launch, and in duty on the liquor and cigarettes sold on the island. Under Major Woods Herm became a miniature tourist centre, and he set up a hotel on the island. At first he lost a certain amount of money, but he persevered, his first aim being to establish a permanent community. 'I looked for happily married couples, preferably with children [he had six children himself] and anxious to bring up their children in the Christian way of life. We now have a school and we have refurbished the little church and have services every Sunday.'

In selecting his fellow inhabitants he had to choose wisely, avoiding those solely attracted by the thought of beer at 5p a pint and income tax at 20p in the pound. Herm is cheap in the sense that there are few temptations to spend, but most essentials are expensive. He had to choose people who would accept him as 'ruler' and leader, and who would be ready to work from dawn to dusk during the season and still work fairly hard in the off season. Many of the people doubled on their jobs. The man who ran the gift shop was also the potter, while the barman at the island's single pub, the Mermaid, was also the island's handyman. There was never any shortage of people anxious to go to Herm to live: the slightest publicity about the island brought many letters from would-be escapists.

By 1960 Major Woods was employing forty-five people and the island's turnover was £70,000 a year. This included the hotel, the Mermaid Inn (licensing hours 10.30am to 11pm), a shop selling gifts made from seashells, boat rides, Herm postage stamps, pottery and the export of milk. A success story, indeed!

When the 1931 census was taken in Britain, it was reported that there were two pairs of Adams and Eves. Each couple had one of the Channel Islands to themselves, one pair in Jethou (44 acres), the other in Lihou (38 acres). The story of Jethou (see p33) also illustrates the rising cost of island purchases.

Sir Compton Mackenzie rented it as well as Herm—at £100 per annum from 1920 to 1934. Then an American, Harold Fortington, paid £2,020 for the lease, which he held until 1938 and surrendered because of the threat of war. Lt-Col and Mrs W. G. Widdicombe took over Jethou in 1948 and used it as a private residence until 1955, when Mr Philip Watkins, a London businessman, acquired it. In 1957 Mr Herman Stockley paid over £8,000 for the lease, which two years later cost £10,000 when Group-Capt Hedley Cliff opened Jethou to the public as a holiday resort, installing a bar and gift shop. He also brought 5 acres into cultivation to grow daffodils for export to the United Kingdom. In 1964 Mrs Susan Faed paid £25,000 for the lease, offsetting the cost of running the island by earning money from day visitors—more than 5,700 paid 45p each for the privilege of landing on Jethou each year. Mrs Faed put the island on the market in 1971 at an asking price of £45,000 for the lease, which included furnishings, the manor house, a café and a motor launch, but the rent remains at £100 per annum to the Crown. So the lease is now $22\frac{1}{4}$ times what it was in 1934.

Lihou is an island that I personally regret, for I failed to buy it when I had the chance before World War II. It was advertised for sale in *The Times* and the lease, which was then a mere £5, was acquired by Capt W. F. Dekker-Webb. After the war Lihou, which lies off the west coast of Guernsey, was bought by Francis Coniff, a horticulturalist, who became joint lessee with a friend. Before the war the island had houses and a factory for making iodine from seaweed, but the Germans used its building for gunnery practice during their occupation of the Channel Islands and the place was ruined.

In 1958 Lt-Col Patrick Wootton took over Lihou and launched his Lihou Youth Project. As the island at low tide is linked to Guernsey by a narrow causeway, communication is easy, and Lt-Col Wootton's plan was 'to train selected young men and women as potential social and youth leaders and to provide a constructive pattern for living in the 20th century'. He does this by organising special fortnightly camps during the summer, each attended by between twenty and thirty young people aged from seventeen to twenty. The Youth Project includes studies of archaeology, marine biology and meteorology.

THE SCOTTISH ISLANDS

At first sight the Scottish coastline seems to offer a wider choice of islands than anywhere else in the British Isles, yet the display of intriguing dots and blobs on the map is deceptive and most of these islands can be completely ruled out for homemaking. Several people have tried, but most of them lost heavily, both in money and spirit.

Lord Leverhulme nearly broke his heart over Lewis and Harris, and, after he had taken great pains to make them pay, they were sold by auction in 1925. He was one of several who have tried to save the Hebrides from depopulation and ruin. It is said that he spent more than £1 million upon schemes for developing Lewis and its adjacent islands, yet after his death one estate of 56,000 acres was sold for only £500. His plan was to make Lewis a great centre for the fishing industry, and he told people: 'Acre for acre the sea around you contains twenty times more food than your land can produce'. But success eluded him and in September 1923 he summoned a meeting in Stornoway and told the people: 'I give you the whole island of Lewis'. Yet the people actually *refused the gift,* though Stornoway accepted the stately castle and its grounds. After Lord Leverhulme's death, some of the most romantic and attractive bargains in islands ever came under the auctioneer's hammer; they included a whaling station, the eerie castle where Barrie wrote *Mary Rose* and deer forests teeming with deer that had not been hunted for years.

Lord Leverhulme's efforts proved that these islands are too remote to be tackled singlehanded, and they offer such a scant existence that they will not support many inhabitants. In the Outer Hebrides even the tiniest island would present a formidable problem for two people trying to exist there. The experiment that Mr Lockley successfully tried at Skokholm could not be repeated in the Western Isles of the Outer Hebrides. You could not hope to make a home here unless you had the co-operation of the inhabitants, or, if your island was uninhabited, of the few people whom you would have to take with you. Signals for help would take hours to answer, perhaps even days, if indeed they were answered at all, and islanders have told me that help is frequently wanted owing to the ravages of severe storms and delays in the arrival of supplies in winter.

The natives can be an obstinate unhelpful people if they do not like you, and their hostility to the outsider is sometimes difficult to overcome. While they show friendliness to the occasional visitor, their attitude is very different if they think the visitor has come to live among them. It helps, of course, if you speak Gaelic, which is the only language some of them know. The English that some speak is difficult to understand. Their obduracy often blinds them to the friendliness of outsiders, so that they seem ungrateful. There is some excuse for this attitude in that for years they have been neglected by officialdom, but they show little desire to help themselves and in many ways have only themselves to thank that the Western Isles have failed to play a more useful role in the Scottish economy. They hate to be managed or to have their affairs run for them, an attitude that defeated Lord Leverhulme, yet they will do nothing on their own account.

Do not think this is a personal prejudice, for, as a Scot, I should like to be enthusiastic about these islands. The combination of native stubbornness and official lack of concern about their future has brought the islands to their present pass. In 1935 Mr T. B. W. Ramsay, MP for the Western Isles, in an appeal to the Prime Minister, spoke of the terrible conditions that existed in the Outer Hebrides, urging them to be classified as distressed areas.

> In almost every Scottish social gathering you will hear *The Road to the Isles* sung or played. But who knows about the roads in the Isles, or the absence of such roads? The dead cannot be taken to their last resting place in a cart, leave alone a hearse. Cases requiring urgent treatment at hospital have not the humane, steady and speedy means of transport which are essential. The poor crofters, their wives and families carry fertilisers, peats and provisions on their backs like beasts of burden.
>
> In a very large number of districts no piers, jetties or boatslips are to be seen for miles. The poor fishermen have to tie their boats to boulders, thereby exposing them to the dangers of the tempest, or they have to undergo the double labour of pulling them ashore and pushing them off with wet bodies.

Nor has the situation changed much since World War II. Once again the people are drifting away from the isles so fast that anyone stepping into their places should think twice about doing better. There is, in fact, evidence of a pigheaded resentment not only among the native islanders but many Scots from the mainland that any outsider should be given the chance of doing any better. This crass stupidity was even vented in the British Parliament in 1958 when Mr Woodburn, the Member for Clackmannan and East Stirling, asked the Secretary of State for Scotland to what extent, in the public sale of islands in Scotland, there were controls exercised by the Government to see that they did not fall into the ownership of persons or institutions inimical to the public interest.

The Minister replied that the 'Government have no powers to prevent anyone from buying an island or other heritable property in Scotland, but there are, of course, various powers under which activities which are against the public interest can be prevented'. Mr Woodburn then asked if it was not a sad commentary on the present conditions of Scotland that so many people were wanting to sell off chunks of Scotland to foreigners. 'Is it not a dangerous thing,' he asked, 'that anyone from America or any other country can come and buy an island and it is difficult for the Government to check what is happening? Surely the Government can find some proper use for them when these islands are not used by their owners?'

The second question really contained the bite in this particular political dog-in-the-manger exchange: the implied suggestion that Crusoeism should be counteracted by nationalisation of islands.

But, having given these warnings about Scottish islands, let it be admitted that there is a brighter outlook in some areas and that some people at least have succeeded in establishing island homes. One advantage is that you will be on a buyer's market and prices and rents are much lower than elsewhere in the United Kingdom. Many Scottish islands have been on the market in recent years and sales and lease offers are continuing. The 600 acre island of Erraid, once the home of Robert Louis Stevenson, was on offer for £50,000, which included nine furnished cottages. Somewhat less was the island of Inch Kenneth, which, in 1965, was offered at £35,000 by Miss Jessica Mitford. With it went a ten-roomed Victorian house.

Further down the scale, Scalpay, an island near Skye, was bought for £25,000 a few years ago, the price covering six houses, grouse, deer, trout and a pair of nesting eagles. Three islands, 5 miles off the north-west coast of Scotland and 16 miles from Ullapool—Carn Iar, Carn Beag and Carn Deas—were offered at £10,000 for the three or £5,250, £4,750 and £1,850 respectively. Although Lord Runciman put up the Island of Eigg for £100,000 in 1966, there are still such remarkable bargains as the 7 acre island of Sgeirean Glasg, one of the delightfully named Summer Isles, which was picked up for £50 at a sale by Mrs I. Mason of Edinburgh, who had the charming idea of buying it for her children. There are also various small islands in the far north and north-west that can be acquired for about £1,000 or rented for £50 a year.

Possibly what stirred Mr Woodburn's chauvinistic wrath was the fact that Stroma, off the coast of Caithness, was once to be presented as a prize in an American television programme. The idea was abandoned after protests and awesome comments by Lord Lyon King of Arms, and, after the fuss had subsided, Mr James Simpson of Moy, Caithness, bought it.

Fishing and shooting are good in most of the Western Isles. It might be possible to develop some of these activities close to the mainland for tourist purposes, but considerable expense would be involved. These islands could provide a permanent residence for either a very wealthy man or one who is prepared to rough it. The low prices can be very deceptive.

Compton Mackenzie, that prince of island-lovers, succeeded for a time in making a pleasant island home in the Hebrides. He built a house on Barra, population 2,250, in the south of the Outer Hebrides. His writings and broadcasts on Barra made it sound delightful indeed, but it must be remembered that Mackenzie is an indomitable island-lover, a Scot, which is more than half the battle when living among the Hebrideans, and that he could when he wished 'escape to civilisation'. Eventually he hold his home, Suideachan, in which he had thousands of books and a large number of gramophone records.

There is one small tip we can usefully take from Compton Mackenzie's experiment. He immortalised the view from his island sitting-room by encasing the long low window with a gilt frame that preserved it as though it were some master's painting. Beverley Nichols in his book about a country cottage, *Down the Garden Path*, deals with the subject of creating a home in the country as though it were an adventure, an attitude even more important on an island, where a dream house is essential to the plan. How many views have you longed to have enshrined before your eyes for the rest of your days? On an island your best room should have such a view, and it should be heightened by every artifice you can employ.

The Hebridean life appeals to that whimsy in the Scots character which none of their writers, happily, seems to lose. You find it in Barrie and in Stevenson, in the novels of Crockett and the philosophy of Hume as much as in the blather of Burns. It even peeps out of the Puritanism of the dourest Scot. The Hebrideans have it in good measure. They are moody, often sour, glowering with righteousness on occasions and on others sinking back into hopeless abandonment to the powers of evil that are so prevalent in their thoughts. Yet the poetry, the magic of romance are never far beneath the surface. They have their moments of gaiety: on Barra and many other Hebridean islands they often sing and dance in the early hours of the morning, and—whimsical touch again—they believe in fairies. There are islands dedicated to the fairies and one—Pigmies' Island—named after them. Fairy stones and fairy rings abound. Compton Mackenzie even had fairy mounds in his garden!

The islanders' chief recreation is the *ceilidh*, a smoking song and dance affair at which whisky is consumed in considerable quantities. Whisky seems to be the one luxury on which the islanders insist. My own impression is that they make it themselves for home consumption, for I cannot believe they can afford to buy as much as they seem to drink. Wherever it comes from, the whisky is drinkable, if fiery. After a few glasses the talk inevitably gets around to fairies, and after several glasses the islanders will even tell you of fairy weddings they have witnessed and ghosts (always green) they have watched.

More attractive propositions in island homes can be found further south and closer to the mainland. The isles of Mull, Islay and Jura and those off the Argyll coast offer a less isolated existence, with good boat services to and from the mainland. They are less rugged and more sylvan than those further north. The Firth of Clyde can be ruled out: its tourist invasion in summer has spoiled such places as Arran. But Skye remains beautiful and unspoiled—a work of Nature's art that is as appealing now as when Dr Johnson discovered its charms. If you love the Western Isles, you might usefully take Skye for your first experiment in living there; for, if you do not fall in love with Skye, the islands are not for you. Nowhere else are there so many different shades of green. When these ever-changing tints appear through the mists as the sun suddenly lights up on a dull day, then you see the island at its best. Between spring and early summer every part of it seems to have a rainbow of its own, the dells are like fairy gardens and the whole scene is lit up like some theatrical drop-piece under the limelight of a kaleidoscopic lamp. Its red hills catch the sun's rays and throw them back at you. No sunsets can compare with Skye's except those in southern Morocco and on the edges of the Sahara Desert. In Skye you can live in a degree of comfort unknown in the Outer Hebrides, or you can live more cheaply but more modestly in one of the tiny islands adjacent to it. Skye has rich agricultural land that compares favourably with any on the mainland.

The whole island is full of relics of the '45 Rebellion and of the Young Pretender himself. At Dunvegan Castle you will find a lock of his hair in a glass case as well as letters from Dr Johnson and Sir Walter Scott. Everywhere you will see some indication of the romance the name Flora MacDonald still conjures up.

Staffa should certainly be visited to get the dramatic atmosphere of the isles. Here is one of the most remarkable of sea organs, the inspiration for Mendelssohn's *Hebridean Overture*. As you enter Fingal's Cave, its fluted columns echo the diapason of the waves as they wash in and out. Then there is Erraid, off the south of Mull, from which at low tide you can wade across the water to Mull itself. This island, on which Stevenson marooned David Balfour in *Kidnapped*, is now a favourite haunt for visitors attracted by its translucent green sea. Near by is Iona, where St Columba preached the Christian Gospel, an island 3 miles long and 1 mile in breadth. It has an eighteen-hole golf course and two hotels, but its inhabitants are mainly crofters.

No one in the past quarter of a century has made a better job of island-owning than Lt-Col Sir James Horlick, who in 1944 bought the island of Gigha (see p34) and became its laird. In 1966, on his eightieth birthday, the islanders presented him with a revolving summerhouse inscribed as 'a testimonial of their affection from the islanders of Gigha'. It was a testimonial justly earned, for Sir James brought prosperity to Gigha, more than quadrupling the island's milk production, setting up a cheese factory, and improving the pier facilities by installing three 10,000-gallon tanks so that the islanders could buy fuel oil at low cost. He has modernised the cottages on Gigha and built more, and created a garden of rare plants and flowers, which he has made over to the National Trust for Scotland.

Oddly enough you are more likely to be able to rent or buy a Scottish island in London than in Glasgow or Edinburgh. The leading London estate agents are by far the best people to approach, though inquiries on the spot can sometimes prove cheaper in the long run.

Gavin Maxwell Enterprises, Ltd, originally formed to look after the taxes of the late Gavin Maxwell, rents out two converted lighthouse cottages on its own islands between Skye and the mainland: Kyleakin lighthouse has been converted to provide three bedrooms and a 40ft long sitting-room that looks up Loch Duich to the Five Sisters of Kintail, and Isle Ornsay lighthouse has four bedrooms and a 27ft sitting-room looking up the Sound of Sleat to the north-east.

Many of the islands have changed hands several times since World War II. In 1944 Mr H. A. Andreae, the banker and yachtsman, bought from the Cathcart trustees the three islands of South Uist, Benbecula and Eriskay in the Outer Hebrides. The price was not then disclosed, but the islands cover about 90,000 acres and include fifty-seven townships comprising some 1,200 crofts. It was stated at the time that Mr Andreae had acquired them for fishing and shooting. Then in 1960 South Uist was sold for £60,000 and North Uist went for the same price. 'We had more than one hundred inquiries for each island,' Mr James Wilson, of the negotiating company of surveyors, stated in Edinburgh. 'The majority of inquiries were merely curious, but a number visited the islands and made offers.'

Just how difficult it is to reach some of the islands may be gathered from these printed instructions of how to get to Taransay, off the west coast of Harris in the Outer Hebrides: 'Tarbert to Horgabust—12 miles. Taxi. Try to hire the Rev Macdonald's rowing boat. If he won't lend it, light bonfire on the headland and hope that the Campbells will pick you up.'

Roderick Campbell recently sold Taransay together with all the stock to John Mackay, who has a garage across the Sound at Horgabust. There are only three families on the entire island, compared with sixteen before the war, and they live by lobster fishing, looking after the sheep and cattle and growing barley, rye and cabbages. Because there is no lime on the island the houses are built of dry stone and the spaces between filled with earth. The walls are anything up to 6ft thick and the roof beams are made of driftwood. The roofs are thatched with bent grass, kept in place by a net weighted with stones—primitive but effective.

Really hardy island-lovers are not deterred by conditions. In the far north, in Scapa Flow, is the island of Fara, only 200 acres. Its working population is now reduced to two—the owner, Mr Watters, and his wife, Ina. The former fishes for lobsters, but working a boat alone and lifting the pots is tough going for a man aged sixty-two, however fit he may be; he has three boats—a rowing boat, one with an outboard motor and a third with a marine-engine. There are three fields under cultivation—oats, cabbages, potatoes and turnips. With fertilisers and modern methods of agriculture the island might support three farms instead of one as at present, with many more sheep.

Mrs Watters is also the postmistress. The other two inhabitants of Fara are, surprisingly in so bleak a place, two middleaged ladies, Miss Ida Woodhams and Miss Meg Peckham, who arrived here some years ago barefooted and with nothing but the clothes they wore. Until then they had lived by sharpening knives, transporting a knife-grinding machine all over Britain. With true island hospitality Mr Watters let them have a ruined house, which they made habitable themselves. They both live on National Assistance, but, nevertheless, they have helped out on the island in many ways. Miss Woodhams can sharpen axes and ploughshares and has even built a winch to haul the dinghy up from the beach.

In 1962 the National Trust for Scotland advertised for two men to settle with their families on Fair Isle, Britain's most northerly island, north of the Shetlands and owned by the Trust. There were then forty-two people on the island and the Trust was anxious that the population should not fall below that figure. The idea was that the men would take up crofting, fishing and weaving and that the women would learn Fair Isle knitting, the island's principal industry. More than fifty families applied.

If you are really keen and can face the hardships, it is possible to survive—but it is also possible to founder disastrously.

3 THE GREAT BARRIER REEF

TAKE a look at the average map of Australia and you will notice a thin whiskery line that curls down the coast of Queensland. That line is the Great Barrier Reef. Turn to a large-scale map of the area, and you will see at once why map research is so important in the pastime of island-hunting. On the small-scale map you see the Reef as a vague, jagged and seemingly uninhabitable line of coral. In reality it represents an area of 80,000sq miles of warm tropical seas dotted with thousands of pine-covered islands. Here exists the nearest approach to an island-lover's paradise.

The Great Barrier Reef offers almost everything—warmth, sunshine, shimmering lagoons that turn from turquoise to emerald, fishing and sunbathing. It is remote and yet close to civilisation, for you can visit these islands without feeling that you are cutting yourself off from the civilised world—the mainland is within easy reach, in some instances only a few miles distant. The outer edge of the Reef is 20-30 miles from the coast on average, and 150 miles at the farthest.

Between the Queensland coast and the coral barrier flanking it is Australia's Grand Canal, an island-studded channel 1,250 miles long. During World War II it was the normal route for coast-bound convoys sailing north. From the air you have a magnificent view of what seems like a tropical Arcady, the smooth sheeny-surfaced sea patterned here and there with a natural mosaic of coral and splashes of coloured vegetation.

The population of these islands has never been great, and only about forty of them are inhabited at the moment. In the past these island-dwellers have been mostly shepherds, goat-

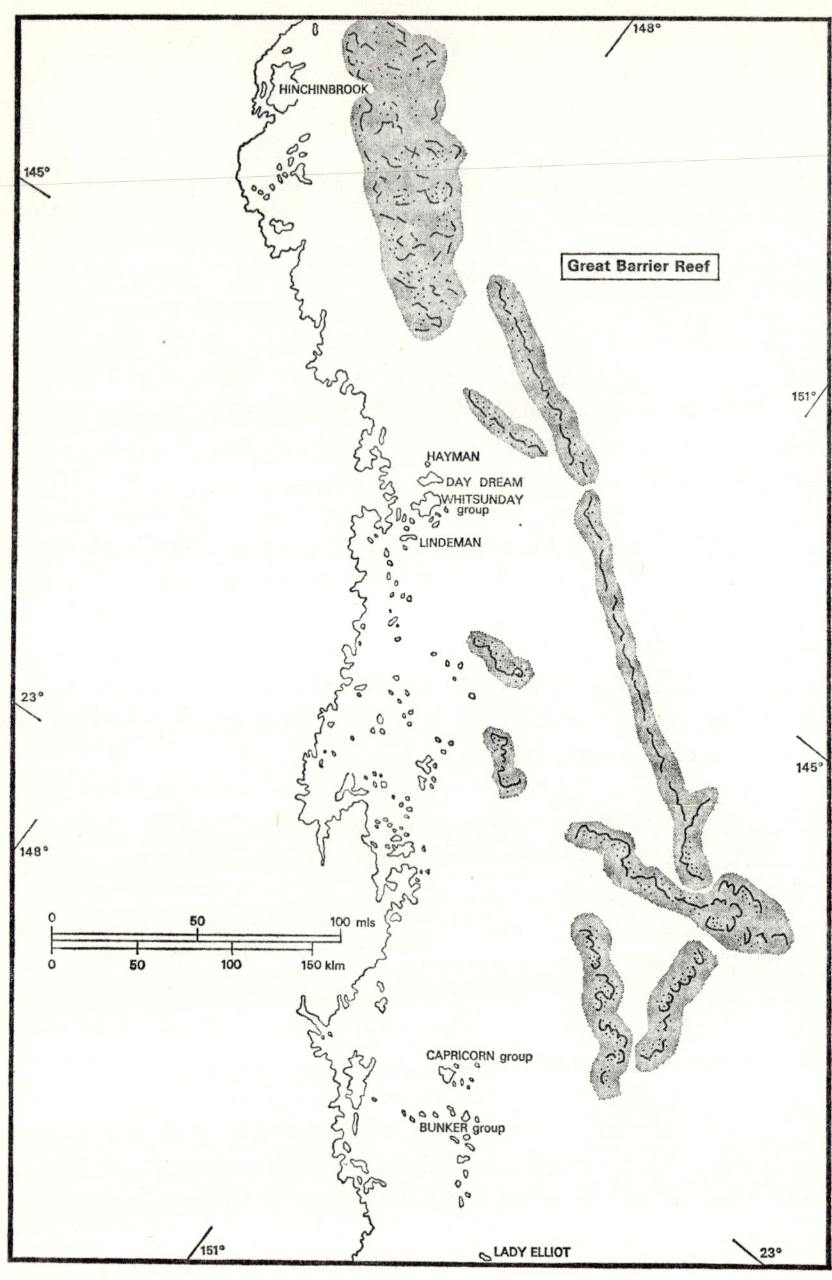

HINCHINBROOK
148°
145°
Great Barrier Reef
151°
HAYMAN
DAY DREAM
WHITSUNDAY
group
LINDEMAN
23°
145°
148°
0
50
100 mls
0
50
100
150 klm
CAPRICORN group
BUNKER group
151°
LADY ELLIOT
23°

keepers and fishermen, but there has always been a small number of escapists among them. On some islands there is excellent grazing land and fishermen have found that by cruising regularly round the Reef they can always make a living, situated as they are close to the mainland ports. For years this small scattered community of shepherds, fishermen and escapists has got along happily, each running errands for his neighbours, passing messages to coastal towns and helping each other in time of need. They liked their solitary existence and it was with a certain amount of resentment at first that they received newcomers who had suddenly discovered this haven.

Some years before World War II a number of enterprising families decided to settle on the islands. In most cases they had come here for a holiday, been entranced with what they saw and fired with enthusiasm for an island life. They put up huts and shanties, choosing islands where there was already a water supply and decent anchorages. The huts were let to paying visitors who came out each season from the large Australian cities. It all worked out very well. Father was usually the handyman about the island, who brought the visitors over from the mainland in his boat, took them for trips around the Reef and did some fishing. Mother did the cooking and cleaning, and looked after the visitors, while the children ran errands to Brisbane and helped generally. Nobody made a fortune, but everybody seemed thoroughly happy, except the old-school islanders, who shook their heads at the holidaymaker invasion.

Australia House, London, on being asked whether there were any facilities for buying one of the Barrier Reef islands, came back with the unexpected answer: 'Why, no, you can't buy an island there, but you sure can rent or lease one. Yes, sir, at one time you could rent an island for as little as £3 a year...' I almost begged the official spokesman to put my name down for an island right away. '... but that was twenty years ago. Today, you can still rent or lease a very small island very reasonably. Possibly for as little as £20 or £30 a year, if you just want a beach and don't mind if there is no artesian well. But for a decent sized island you might have to pay £4,000 for a ten years' lease, or for a really large island as much as £10,000 for fifteen years' lease.'

Further inquiry established that the Barrier Reef islands are nominally Crown land and the privilege of bidding for them has to be carried out through the Queensland Government in Brisbane, in whose sandstone-pillared Lands Office may be found documentary evidence that Britons retain their dream of romantic South Sea islands and adventure despite the distance they have to travel to reach them. The clerks can show you letters from London, Birmingham, Glasgow and Manchester, and in the last ten years inquiries have been coming in from San Francisco, Los Angeles, New York, Seattle and Detroit.

Of course this demand does tend to push up the rents and lease prices, the former being based on a percentage of the capital values of the islands. There is one system by which you can obtain an island on what is euphemistically described out there as the 'ever or never' system. You would need Australian legal advice before entering into any such arrangement, but it is similar to buying a house in Britain under a solicitor's loan and having only to pay the interest on the capital each year. If you bid, say, £1,000 ($2,400) for an island, you pay interest of about 5 per cent per annum and nothing else.

But in recent years the Queensland Government has become conscious of the popularity of the Barrier Reef and its commercial possibilities. The waiting list of applicants has run into hundreds, and, while applications for leases are still received and considered, their granting depends very much on what the applicant proposes to do. The previous essential qualification that applicants should be British subjects has been waived on a number of occasions, particularly if the applicant's ideas on development seem desirable. The

Page 51
Lindeman Island in the Great Barrier Reef

Page 52
Looking towards the mainland of Australia from Lindeman Island

larger the island the more detailed plans does the Government expect.

Until now there have been no bylaws or regulations to worry about, the island tenant merely being expected to conform to normal standards of sanitation and hygiene. Some of the islands possess artesian wells, which solve the problem of fresh water supplies, and in many of them you can grow vegetables and fruit to a limited extent, as well as keeping sheep and goats. Whales and turtles play about in the blue waters and the multi-coloured reef shelters rainbow-tinted tropical fish. Mackerel, schnapper, tuna, bonito, coral trout, carbs, gropers, gamefish, swordfish and lobsters can be caught in this fisherman's paradise, and for a novelty you can hunt turtles with a spear.

On the north the Roof extends almost to New Guinea, and on the south it reaches below the Tropic of Capricorn to Breaksea Spit. Between the Spit and Swain Reefs there is a scattered assortment of reefs and coral islands. The most southerly coral island is Lady Elliot Island, which is about ½ a mile in circumference and has a lighthouse. Nearby are the Bunker and Capricorn groups, separated from the Swain Reefs by a 60 mile channel. In the Coral Sea the Barrier fans out into various unconnected reefs and islands that extend through Torres Strait to the west along the coast of New Guinea, where they disappear altogether.

Some islands are mere mounds of white sand, others are profusely covered with vegetation, pisonia scrub, pandanus palms and flowers. It is a delightful experience to drift over the living coral in a glass-bottomed boat and see one of the most remarkable and beautiful spectacles in the world. In whorls, tendrils, branches and lumps the coral presents a living kaleidoscope of prismatic effects. Coloured fish dart among its green canyons and fairy grottoes, looking like birds of many hues.

On the inner edge of the reef are the 'high' islands, so named to distinguish them from the 'low' islands of the reef proper. The former, which are rocky and thickly grown with vegetation, are the tops of a mountain range submerged ages ago when the sea rushed in behind the coral barrier. They are not coral islands, but the active polyps have fringed them with reefs, though the coral of these reefs is usually far less prolific than, and the land vegetation often quite different from, that found in coral cays. Largest of the islands is Hinchinbrook, which is almost 30 miles long. Its peaks, towering and mysterious, rise up more than 3,000ft above the blue waters. It is far and away the most impressive of the inshore group, yet somehow strangely remote—there is no doubt that a large uninhabited island can have a strange effect upon one, and that smaller islands are in some way friendlier.

Names have a lot to do with an island's appeal. Who could resist Day Dream Island in the Whitsunday group, the most popular islands with tourists? It has been acquired by Mr Reginald M. Ansett, Australia's most enterprising tourist manager. He was born on Friday, 13 February 1909, a fact that seems to have imbued him with a determination to defy superstition and take risks. Certainly fortune has favoured this man, who left school at fourteen, learned to fly before he was twenty-one, and at twenty-two owned and drove a bus that ran between Hamilton and Ballarat, Victoria. Now he owns an airline, hotels, a bus-building factory and certain Barrier Reef Islands, including Day Dream.

He more than anyone has managed to interest the Queensland Government in tourist development of the islands. He planned Barrier Reef holidays, using motor yachts and putting up modern hotels on some islands. The subsidiary of Ansett Transport Industries Ltd is the Barrier Reef Islands Ltd. This new company has taken over Hayman and Day Dream islands in the Whitsunday group and fitted out a luxury motor cruiser to take tourists along Australia's 'Grand Canal'.

Another key figure in these development plans was Squadron Leader Stewart C. Middlemiss, a former insurance man who became head of the Barrier Reef Airways. He ran a regular service from Brisbane to Heron Island, one of the Capricorn group, and a weekly trip to Lindeman, Day Dream and Hayman, which are 600 miles north of Brisbane. The private enterprise of Middlemiss and Ansett inspired the Queensland Tourist Development Board to make a report on Barrier Reef possibilities. The Board now tends to withhold island leases from shanty settlers in favour of developers —and the Australians are not the only people who see the commercial possibilities. British and Americans have made tentative inquiries for investment. Parties of Americans have surveyed the islands and the famous American game-fisherman, Kip Farrington, declared: 'This is the world's finest paradise'. But commercial development is controlled, for many Barrier Reef islands have been declared national parks.

Those who have ideas about leasing an island off Queensland should first decide what they want to do there. If you have capital to spare, and are prepared to set up a modest guesthouse, for instance, then the chances of your plans being approved will be greatly enhanced. The Queensland Government wants settlers rather than escapists, which is sad in a way, but there is the consolation that you are unlikely to be duped into renting a useless island. If your application is granted, the odds are that you will get an island where you can at least make a living of sorts.

As the exploits of Ansett and Middlemiss show, the Government is wide awake to all sorts of ideas and the possibilities for island home-hunters in the Barrier Reef are limitless. Tourism will undoubtedly provide scope for doctors, dentists, technicians and members of other professions. It is interesting to note that the Tourist Development Board envisage such things as 'a brighter night life for the islands, cabarets, floor shows and theatres'. Little has yet been done in this direction, but it would seem to present a warning for those seeking absolute solitude and encouragement for those wanting to take a gamble. To the old inhabitants the idea of cabarets, hotels and aeroplanes is anathema, though there are thousands of islands and a little development here and there is not going to spoil them all. In 80,000 square miles of sea, sand and jungle you may easily escape from night clubs, luxury hotels and pop; and with a little plumbing, a few more amenities, the islands should be even more attractive than they are already.

Perhaps the most famous of the 'Old Guard' of the Barrier Reef was Edmund James Banfield, who wrote that Australian classic, *Confessions of a Beachcomber*. For twenty-five years he lived on Dunk Island, between Townsville and Cairns, and his grave on the island bears an epitaph worth quoting: 'If a man does not keep pace with his companions, perhaps it is because he hears a different drummer. Let him step to the music which he hears'.

Men of many nations have settled here, men of various views and differing background. Most colourful of these was perhaps Sir Winston Churchill's first cousin, Hugh Frewen, brother of sculptress Clare Sheridan. Hugh Frewen skippered a 50-ton ketch, *Viking Ahoy*, which cruised in and out of Bowen, a small port on the Queensland coast, and knew the Barrier Reef better than anyone. He delved into the ancient stories of the Reef, checked up on tales of treasure trove and entertained his visitors with a hundred romances.

There are some queer creatures in this region, like the walking fish. Best known as mud-skippers, they are ugly looking creatures with large heads and bulbous eyes, and when the mud flats are exposed hundreds of them can be seen clinging to the mangrove roots, or skipping along the sand. Another curious Barrier Reef inhabitant is the dugong, or sea-cow. This, say Australians, is the original mermaid, and many are the tall stories sea captains tell of it. A pig-faced creature, 9ft long and weighing half a ton, it is a mammal which takes to the water but breathes air. It has shoulders, with flippers that resemble human arms, and lives off the seagrass that grows in shallow waters. Dugongs are, technically speaking, *halicore australis.* They have no blonde tresses, not a semblance of a come-hither look, but, seen in the uncertain light of dawn, hugging their young to their breasts, they look almost human. The mothers' heads bent over their young make a strangely moving picture. They have a marked 'family' sense and, if a mate or child dies, the other mate remains in the vicinity for days, whimpering and sighing. The sighings must undoubtedly have stirred some sailors of bygone days who, not having seen a woman for months, became quite delirious at the sound of this strange female heart-cry.

The extraordinary sucker-fish is a useful friend to the ingenious aborigines. It grows to about 4ft, and attaches itself to a turtle, shark or other large creature by means of a 'sucker'—an oval-shaped disc, arranged in overlapped plates like a Venetian blind—at the top of its head. By depressing the base of the disc the sucker-fish frees itself by swimming forward against its host and spreading out the plates. No amount of backward pull will release the sucker. The aborigines use these creatures to catch fish or turtles: sighting a possible catch, they throw a sucker-fish overboard with a line attached to its tail, and once the sucker-fish has fixed itself to the turtle or other fish, all the aborigine fishermen have to do is haul them both in.

The variety of natural phenomena is one of the chief charms of the Barrier Reef, and lovers of the exotically picturesque will delight in the sea anemones. They look like chrysanthemum blooms and cactus dahlias and are called 'flowers of the Reef'. In most places in the world they rarely grow larger than 2-3in across, but here they reach 12in quite commonly, and sometimes 2ft.

One of the most delightful places is Heron Island, 1 mile in circumference, 300 miles north of Brisbane and 45 miles off the Queensland coast. It has an emerald-green lagoon, enclosed at low tide by a coral rampart, glistening white beaches, a jungle of 60ft pisonias, with tournefortias and casuarinas, and good deep-sea fishing. From the Barrier Reef

Airways plane it looks like a spray of foliage floating in liquid light and the coral gardens at the bottom of the lagoon seem to be only a few inches below the surface. Actually the lagoon is 10ft deep, ample to take a flying boat. Crossing it is like gliding over a clear glass bowl underneath which all the secrets of the sea are visible. Its surface takes on a hundred tints, mauve here, purple there, pinks fading into violets and vivid orange melting into deepest green.

A motorboat takes visitors to neighbouring islands and reefs. Mutton birds (wedge-tailed sheerwaters) use the island as a nesting place. Turtles are common, as, too, are the giant clams that are found all along the Reef. The latter show a variety of design and colouring that is unequalled elsewhere. They rest on the bottom, hinge side downwards, and present gaping jaws to the surface, absorbing the microscopic forms of life on which they live. The fleshy mantles that line the shell are brightly coloured, but no two have the same colour pattern—one may see royal purple streaked with thin green lines, brown and ochre bands, deep blue blotched with dark green, or an opalescent green lined at the edges with white and covered with small black spots. These piscatorial beauties are commonplace off Heron Island, but it takes weeks to explore all its less obvious wonders.

Barrier Reef Airways call at Lindeman Island (see p51) once a week, and this compact territory, only 3sq miles, provides variety of another kind, including the 712ft peak of Mount Olden, and a landscape of bluff headlands, grassy slopes, forest, savannah and dense tropical jungle. More than seventy similar islands are grouped around Lindeman. Nearby is Day Dream Island, already referred to, which is shown as West Molle on Admiralty charts; it is 65 acres of sickle-shaped hilly country, 1 mile long and possessing the most marvellous beaches.

If you stay at either Lindeman or Day Dream, you can stroll through cool bird-haunted jungle. When the southern states of Australia are getting their worst weather, these tropical skies are eternally blue. The only bad weather months on the Whitsunday group are February and March, which form the cyclone season—a succession of storms and unending days of steaming rain. Most of the islands are not open for visitors during these months, but the off-season is short and causes the minimum of inconvenience to the inhabitants.

When looking for an island cheap, remember that it will be much easier to get if you pay for a whole ten or twenty years' lease. But do not expect much for a cheap rental. You are likely eventually to obtain a very tiny island that may be admirable for a camping holiday or occasional visits, but will need some money spent on it if it is to be brought up to acceptable standards of habitation. Often, as an alternative, several island-seekers have clubbed together to obtain the lease of one large island, a policy that has much to commend it, provided the persons concerned can get along together.

On 23 October 1961 Mr Bernard Stanbury, a thirty-two-year-old Mayfair photographic printer, and his wife Sally, began to look for fifty families who would be prepared to join them in a remote island community 'free from nuclear hazards'. He wrote a total of 564 letters to estate agents and Government departments in search of a suitable island, and thirty-two islands were offered him as a result.

One of Mr Stanbury's letters was to U Thant, the then acting-Secretary-General of the United Nations Organisation, in which he said he hoped to 'establish a farming and fishing community and to be self-supporting', but it so happened that at that time the United Nations Trusteeship Council was engaged in searching for an island on its own account. The islanders of the trusteeship territory of Nauru, in the Central Pacific, were hoping to resettle themselves on another island, since the phosphate resources of Nauru were expected to become exhausted within thirty years. Reluctantly the Trusteeship Council had to turn down both the request from Mr Stanbury and that from the islanders of Nauru. The moral of this would seem to be that there is probably a very good case to be made out for the establishment of an International Agency of Islands.

Eventually Mr Stanbury narrowed his choice to two islands in the Great Barrier Reef—Middle Percy Island, 7sq miles and costing £16,000 for a ten-year lease; and Hummock Hill, 10sq miles for £12,000. The former, though smaller, had a furnished house and the owner had grown fruit there and established a farm with 1,000 head of sheep. There was also a natural harbour with four boats included in the price. Hummock Hill, though larger, was much bleaker, and had no house, but some 3,000 acres had been cleared for cultivation.

Middle Percy Island was finally selected after a thorough survey had been made. Mr Stanbury had more than 3,500 applicants to join him, and forty of these, from Britain, Holland and Italy, were actually selected. Two of them, Jeff Forse and his wife, made the survey of the island, which was 70 miles off the coast of Queensland. There was a seven-room house, which stood 750ft above sea level, 1½ miles inland, nestling among coconut palms, hibiscus shrubs, mango and banana trees. There were 150 acres of arable land on which maize, vegetables and fruit could be grown and acres of cedar trees from which furniture could be made. The island's beaches abounded with oysters and crabs. Two clear-water streams only 500yd from the house had never been known to run dry, while there were nine huge rainwater tanks near the house. Coffee trees had grown wild to such an extent that many acres of them had had to be burned down each year to keep them from spreading.

It seemed indeed an island holding out every prospect of a practical paradise for the eight married couples and six single men all ready to form the expedition's vanguard, which included a carpenter, farmer, nurse and land-drainage contractor. But money proved the obstacle to this adventure. Six months later Mr Stanbury had to admit that the raising of the necessary cash was a major stumbling block. The dream began to crumble. Possibly the mistake here was in numbers.

To form a community of forty people may not sound a great task, but it is an awkward number to cater for, probably more difficult than if 100 were involved. A much more practical proposition would have been an island costing, say, £8,000 ($19,200) for fifteen people. It is still relatively easy to find islands suitable for not less than six and not more than fifteen persons, and it is below and above those figures that problems begin.

However, while Mr Stanbury may have been too ambitious, it is no use setting out with hope alone and no adequate equipment, as did the Carpenter family from London. In 1961 Mr Carpenter, his wife, two sons and a ten-year-old dog, Perkins, set out from Cooktown, Northern Queensland, to Flinders Island, 140 miles distant. Showing great determination and courage, they sailed across shark-infested seas in an 11ft dinghy with only 8in of freeboard. They had a canteen of water, four bags of provisions and an umbrella. A Cooktown fisherman who saw them go said: 'The dinghy was thirty-eight years old. Carpenter paid a fiver for it and I wouldn't have gone across the bay in it. It was as heavy as lead. It had been waterlogged for six months and when I last saw them it still looked waterlogged. I just couldn't believe they would try to row to Flinders Island in that old tub'.

In due course the family was reported missing and Air Force planes went up to look for them. Eventually the Carpenters were sighted on a deserted beach on Cape York Peninsula 40 miles away. An RAAF Dakota dropped water and food and a message to say a police launch was on its way to them. But Mr Carpenter merely shook his furled umbrella at the plane and then scrawled in the sand: 'I refuse to acknowledge police message. Proceeding north'.

Such is the faith of island adventurers!

4 THE PACIFIC

FRENCH OCEANIA

If it is romance you want, then the Pacific is the place for you. Having said as much, it is perhaps as well to add the rider that you must not merely take your romantic spirit with you but be able to sustain it. This may be difficult, for it is the fashion nowadays to decry the romanticism of Gauguin, Somerset Maugham, Robert Keable, Jack London, Rupert Brooke, Robert Louis Stevenson, Pierre Loti and others who have written about this territory. The sneers come in part from disillusioned writers who lack the ability to express what they have seen, and partly from an attitude of philistine commercialism that says: 'So many nice things have been said about the Pacific that our books won't sell if we say the same, so let's say all the bad things we possibly can'.

It is essential that you should become in tune with your environment. You must be able to feel that the sapphire skies, the azure lagoons, the palms, the flowers and the scrub are companions, that they strike in you a responsive chord. Remember that the Pacific is a long way off, and the further you go the greater becomes your need for something utopian. But Utopia is only achieved by supplying your own brand of magic. In fact you cannot appreciate life on a Pacific island unless you are something of a pantheist. You must feel, with Rupert Brooke, that here is 'an ideal place to live and work in ... that Europe slides from me'. If palm trees become monotonous, if 'views' mean less than people, if the simple things of life bore you, then the Pacific will be completely disillusioning.

The islands of this vast ocean have changed considerably in the past twenty years, but there are still many that answer to the description of our nursery dreams. Most of them are sufficiently removed from civilisation to remain unspoiled, and except for Tahiti and Hawaii and a few others the commercialism of tourism has left them untouched.

I regularly take a magazine with the entrancing title of *The Pacific Islands Monthly*. All who seek Pacific islands ought to buy it, for it contains a wealth of information. Several years ago it published the following notice:

> TROPICAL ISLAND AVAILABLE
>
> An Englishman, Mr Frank Holmes, has decided to retire from his beautiful island of Tupai, situated 120 miles from Papeete and eight miles from Bora-Bora.
>
> With a shimmering blue lagoon (where a seaplane could easily alight), a fine climate, 150,000 coconut trees, tropical fruits, excellent buildings and equipment, the island has everything necessary for an enjoyable and stimulating existence. There is even a reputed hidden treasure to which the owner reserves the right to a percentage if found within the next twenty-year period. Anyone interested should write to Mr Oscar Nordman, Agent, Papeete, Tahiti.

I was intrigued by the advertisement and wrote for information from three sources, two of them completely disinterested. I managed to get photographs of the island that proved it was all it was reputed to be—an island of glamour and beauty, a little kingdom in itself, much more akin to our daydreams than anything Alexander Selkirk encountered.

Tupai is in the Society group of islands, discovered by Captain James Cook in 1769. It possessed all you needed for the ideal home, and its buildings had all modern amenities. Its price then was £185,000 ($444,000), but it was possibly open to bargaining. Tupai was capable of producing 1,000 tons of copra a year, which would be worth around £80,000 ($192,000) a year in gross income. The island was well watered

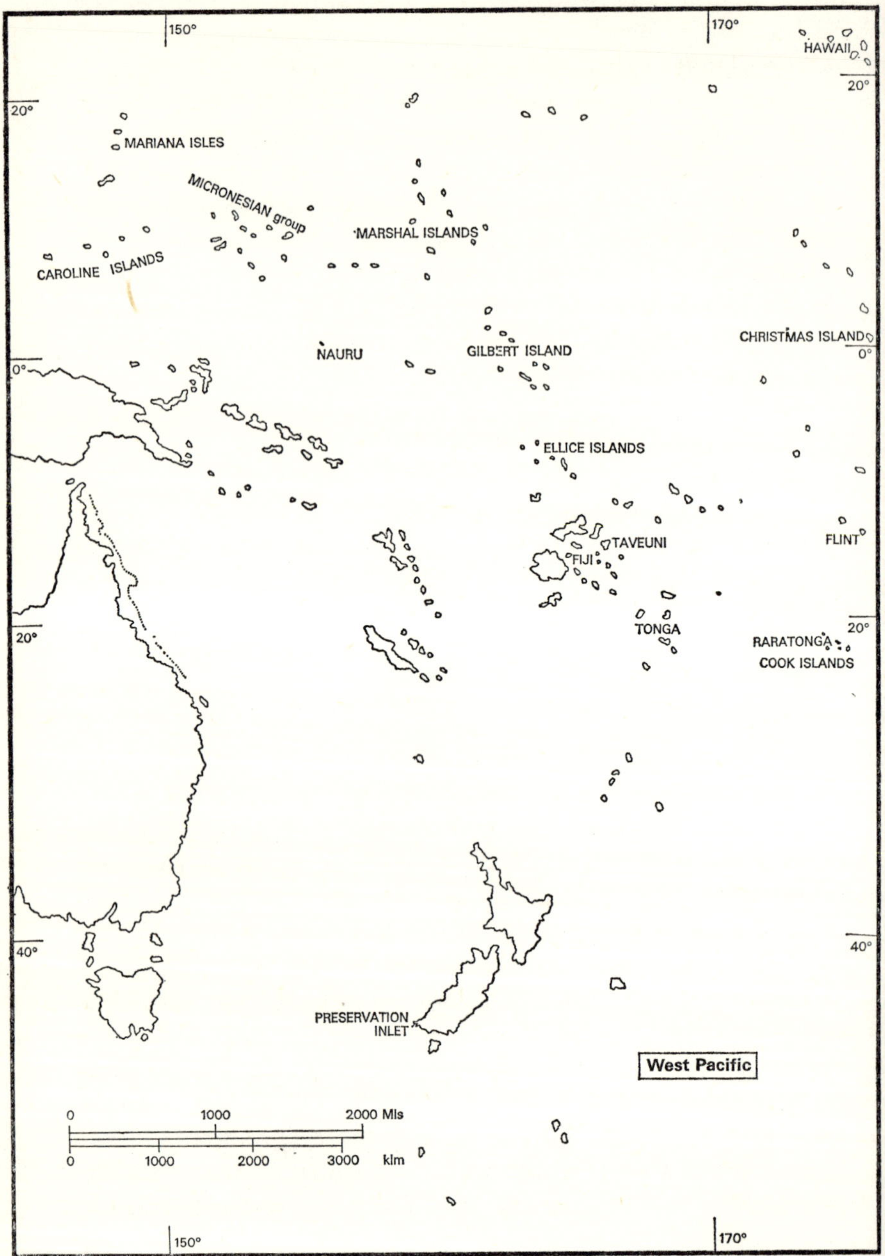
150°
170°
HAWAII
20°
20°
MARIANA ISLES
MICRONESIAN group
MARSHAL ISLANDS
CAROLINE ISLANDS
CHRISTMAS ISLAND
NAURU
GILBERT ISLAND
0°
0°
ELLICE ISLANDS
TAVEUNI
FIJI
FLINT
TONGA
20°
20°
RARATONGA
COOK ISLANDS
40°
40°
PRESERVATION INLET
West Pacific
0
1000
2000 Mls
0
1000
2000
3000
klm
150°
170°

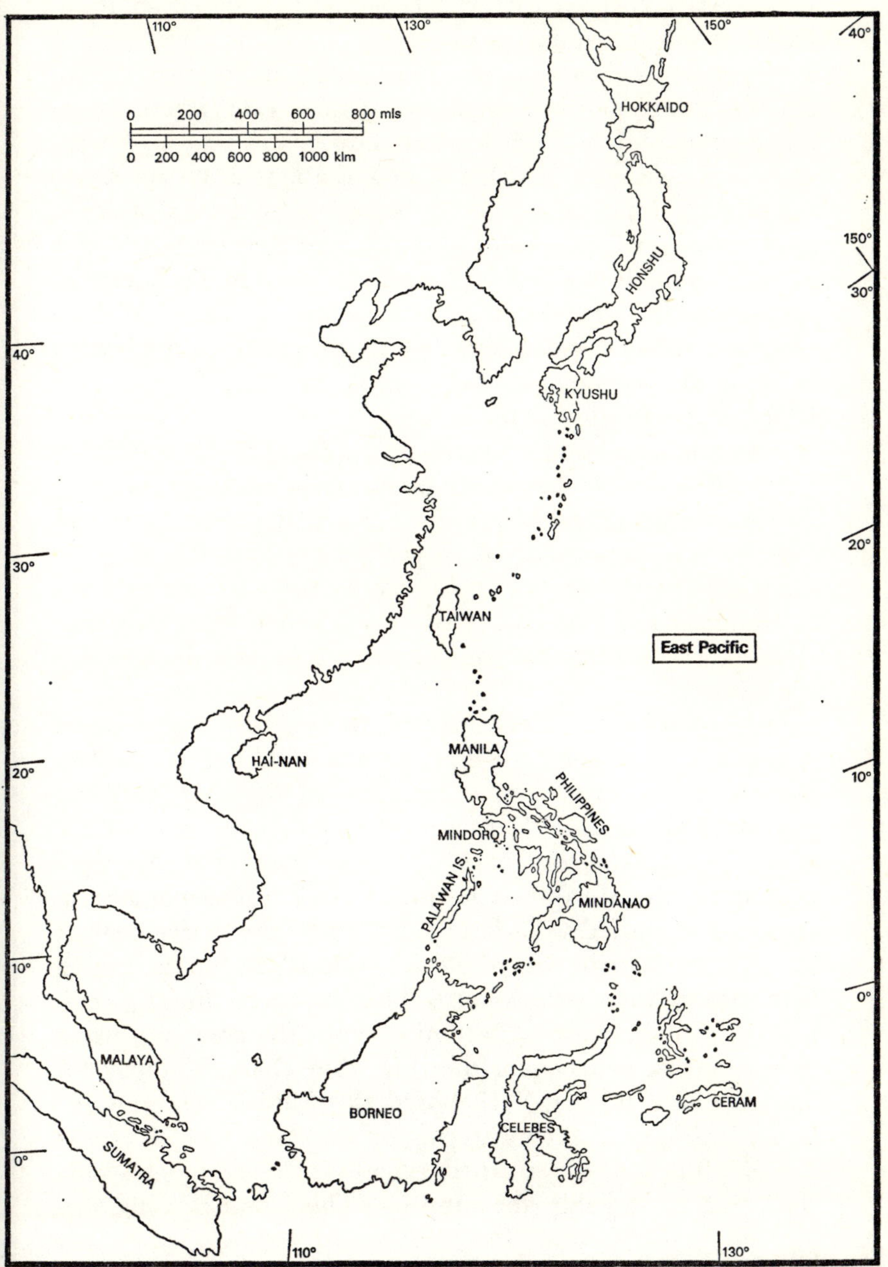

110°
130°
150°
40°
0
200
400
600
800 mls
0
200
400
600
800
1000 klm
HOKKAIDO
HONSHU
150°
30°
40°
KYUSHU
30°
20°
TAIWAN
East Pacific
HAI-NAN
MANILA
20°
10°
PHILIPPINES
MINDORO
PALAWAN IS.
MINDANAO
10°
0°
MALAYA
BORNEO
CELEBES
CERAM
SUMATRA
0°
110°
130°

and wooded, covered with vegetation and possessed an abundance of tropical foliage and flowers. Its climate was that of a perpetual English summer. Fruit trees—fig, orange, citron, papaya, mango and breadfruit—flourished everywhere. Vegetables grew with a little care and vanilla was produced with success. The extent of cultivation was about 2,500 acres, the remainder of the island being occupied by groves of trees, sandbanks, small lakes and an outer reef that offered refuge for innumerable seabirds, fish and turtles. The perimeter of the outer reef was about 30 miles long. A lagoon of deep sheltered water 4 miles long, which provided an unrivalled harbour for hydroplanes operating to and from Papeete, lay in the centre of the island.

There is a story of treasure in Tupai that dates back to 1822, when the crew of a brig *Araueana* of the Chilean Navy mutinied and turned pirate, plundering the coastal cities of Peru. The ship reappeared in the Society Islands under the name of *The Providence* apparently loaded with loot—chests full of gold and gem-encrusted vessels stolen from Peruvian churches. It put into Tupai and here, it is said, the treasure was buried.

Sixty years later there arrived in Tahiti by ship from America two debonair young brothers calling themselves Rorick, who claimed to be members of the then reigning Spanish royal family. They were hospitably received in the island, but did not stay long. They announced that they were leaving for the Marquesas Islands. Details of the momentous voyage that ensued were later supplied by the native cook of the schooner, a man named Neri. While at sea, the two young men showed their true colours. They captured the ship, killing some of the officers. Battening down the crew and using only the cook to navigate the ship, they changed course for Tupai. There, leaving Neri aboard, they landed, accompanied by an intimidated crew carrying an assortment of picks and shovels. The party remained ashore for several days, only returning to the ship for supplies. Then, one evening, the crew were all poisoned and their bodies thrown into the sea.

Following the usual pattern of such marine ruffians, the brothers then changed the name of the ship. Painting the word *Ngamanuariki* on her bows and hoisting the flag of the Cook Islands, they sailed for the Caroline Islands to refit. There Neri escaped, jumping into the sea and swimming through shark-infested waters to safety. He denounced the brothers to the authorities and they were arrested, tried and convicted of mutiny and murder. The younger brother was sentenced to twenty years' imprisonment, the elder to a life sentence. The former lived to be pardoned, but his brother died at Cayenne in 1920. Neri, the cook, died in hospital at Papeete in 1918, talking in his dying hour of the buried treasure at Tupai. His theory was that the brothers found the treasure, but that for some reason or other they did not consider it a favourable time to remove it. It is a story that is still talked about by the inhabitants of the Society Islands. As far as is known no further search for the Tupai treasure has been made since the Rorick brothers' visit to the island, nor have the owners of the property made any systematic attempt to locate it. Modern methods of recovering such treasure trove might prove successful.

Other islands in the Society group, which are all French, come on the market from time to time, but they are smaller islands, cheaper than Tupai. The most romantic of the Society Islands, of course, is Tahiti, but it has changed considerably since World War II. It is not the escapists' paradise it once was, economic circumstances now causing the authorities to discourage beachcombers. Though the French encourage tourism, the bureaucrats who administer Tahiti inflict petty restrictions on tourists and would-be settlers alike. Tahiti today is gagged by red tape. Before World War II, all that was necessary for a visit was a round-trip ticket and a visa from the nearest French consul, the latter obtainable in a matter of minutes. Now a prospective visitor must pay the consul the cost of a cable to Tahiti and its answer from the Government giving permission before the consul can issue a visa. And a close check is kept on the visitor during his stay.

If you want to visit the outer islands, a letter to the Governor is necessary. Before permission is granted the official in charge of such matters carefully consults the bank and, if the visitor is not guilty of trying to live too cheaply—that is, if he is bringing in sufficient money by bureaucratic standards—he is allowed to make the visit. The French authorities apply their strict regulations to protect both the tourist and the Tahitians, and, once you are accepted by officials, everything is made easy for you and life in Tahiti is extremely pleasant. Just as unwelcome as beachcombers are black marketeers in currency.

Tahiti can well claim to be one of the most beautiful islands in the world. Only Papeete, its rather ramshackle capital, spoils it. The green-clad island, rising 7,340ft out of the sea, is slightly larger than the Isle of Wight. Lady Brassey describes her first glimpse of Tahiti as follows: 'a girdle of luxuriant and intensely green tropical vegetation gorgeous with gaily coloured leaves and blossoms and golden-hued fruit, encompasses this delightful harbour, while corals, seaweed, zoophytes and fish of every possible tint are seen as in a wild garden, beneath the transparent waters on which we are floating'. But today the galvanised sheeting of the roofs, the shanty-like buildings spoil the effect somewhat. It is in the interior that Tahiti is most enchanting, in the deep shaded valleys among the bubbling streams and waterfalls. The beautiful flowers include the noble purple of the bougainvillea, gardenias (*tiare Tahiti*), mimosa, hibiscus (which makes such charming decorations for the hair of native girls), convolvulus, yellow oleander and the sweet-smelling frangipani. There are fruits to suit every taste—bananas, plantains, dates, oranges and guavas.

The natives never seem sad. Laughter and smiles are an essential part of their nature, and they have an almost Grecian charm in their physique. They live and work as a team, which

Page 69 (above) *Whitsunday Passage in the Great Barrier Reef;* (below) *The Molle group in the Great Barrier Reef, one of which is delightfully named Day Dream*

Page 70 (above) *Pulau off Labuan;* (below) *Aitutaki, coral isle in the Cook Group*

explains why they seem so carefree and indolent: if there is a job to be done, they do it together quickly and cheerfully, then they sit back, relax and enjoy themselves until another job presents itself. But they do not seek work.

Of all the French group of islands the Marquesas and Bora-Bora are by far the most typically Polynesian and the least spoiled by civilisation. Here you will see children who are lovable and unsophisticated, quite unlike the worldly-wise infants of Papeete. They radiate happiness with their sunny laughter and winning ways. The natives do not beat their children, thinking it sinful for anyone to lay hands on a child, yet infants and teenagers alike are carefree without being wayward. The natives of Bora-Bora—they call it Pora-Pora—are the most attractive of all the islanders. As you pass by they will great you with a smile and the phrase *'Iorana'*, which means, simply, 'My love to you'. In Bora-Bora, too, you will find the genuine Polynesian dances—immeasurably better than the *danses de ventre* and semi-Europeanised dances of Tahiti. They are simple and almost monotonous, yet in a subtle way both stirring and erotic. Karl Baarslag, describing these dances, wrote:

> Each dance seemed to me to be more daring and wilder than the previous one. Before we scarcely apprehended, we were witnessing one of the unforgettable spectacles of the South Seas—the stark, barbaric beauty of the procreative act set to dance rhythm and symbolism without the slightest suggestion of lewdness and offensiveness. One had merely to shut one's eyes to recognise aurally the symbolic slow building-up of the tempo and intensity of passion.

There is, alas, a hardening of attitudes in Polynesia, and growing concern about land speculation has caused some islands to tighten up their immigration laws. It is the Westerner himself who is mostly affected by the changing attitudes. He has brought much of it upon himself in two ways: firstly by following policies, such as land speculation, which invite protective legislation by the islanders, and secondly by giving some of the islanders independent rule. The latter trend will almost inevitably continue. There has been a recent decision by the Cook Islands (see p70) to ban Europeans from citizenship, a ban not just directed against land speculators but against a young London art student who set himself up as a modern Robinson Crusoe on an uninhabited coral atoll some 540 miles east of Pago Pago. The student was named Michael Swift, and he was landed from a trimaran by a New Zealand sailor on Suwarrow Island in September 1965, with tools to build a house, fishing gear and—oddest item of all—40,000 refuel units for his cigarette lighter. Suwarrow Island is technically under the government of the Cook Islands, a New Zealand dependency, and the Islands authorities were quick to claim that Mr Swift was an illegal immigrant and that he had no visa to land on the atoll.

THE GILBERT AND ELLICE ISLANDS, FIJI

To be an out-and-out escapist, you need to be quite sure that you can withstand successfully the hundred and one disadvantages and difficulties that lurk beneath the romantic surface of a Pacific life. Apart from the rigid immigration laws there are the problems of a fairly high cost of living everywhere in the islands, a diet that is often monotonous and sometimes deficient in vital vitamins, trying climatic conditions and a wide range of depressing and horrible diseases. You need in many of the islands not merely an adequate reserve of capital but a tough constitution. Sanitation often leaves much to be desired, and the scourge of the mosquito and other insect pests has not been eliminated even in some of the British possessions. It is no use putting your faith in mosquito nets, a daily dose of quinine—still the most reliable of anti-malarial drugs, despite what the experts may say—and DDT. I have persevered conscientiously with all these remedies, and they did not prevent my developing a chronic form of recurring malaria that took fifteen years to eradicate. You may be told by doctors that the famous fifty-day cure with quinine and drugs will rid you for ever of the worst forms of malaria, but this only applies after a very long period away from the area, and you must stay out of it for the rest of your days. There is nothing so lowering to one's health as a malarial fever that recurs every few months.

There are, however, far worse ailments easily contracted by Europeans even after taking precautions. In certain areas of the Pacific leprosy, filioriosis, typhoid and typhus still take their toll and elephantiasis—that most horrible and disfiguring of diseases—is a particularly deadly and prevalent scourge. Medical attention is not easy to get. Dentists and doctors are scarce and their fees are high. A friend who lived in a Pacific island had to travel 2,000 miles to have his teeth attended to. Lack of medical and hospital facilities, the scarcity of dentists and the almost total absence of specialists of any kind do suggest that doctors, dentists and specialists themselves stand the best chance of surviving and making a living in the Pacific. In many places they would be welcomed immediately and all regulations waived in their favour. One dentist established his headquarters in his yacht and by cruising around several of the islands acquired a practice worth £5,000 a year tax free. The New Zealand Government finds it extremely hard to obtain medical officers willing to accept appointments in its island service and Fiji has also had some difficulties in recruiting medical officers. Anyone interested in trying his luck either as a governmental service officer, or as a private practitioner, stands an excellent chance of success.

Food in the Pacific is not always good. The tourist may fare well in Honolulu, but on the whole there is a lack of fresh meat, and a diet of fish can become very monotonous. Good green vegetables are practically non-existent and those there are lack vitamins, while, strangely enough, the popular idea that Pacific islands are teeming with luscious fruit ready to drop into your lap—as in Tahiti—is just another beachcomber's fallacy. Fruit is actually scarce in many places, even if abundant in some, and only a daily ration of anti-scorbutic limejuice will compensate for its lack. Scurvy is a disease that attacks many Europeans and the best answer to it seems to be the regular drinking of the native tipple—toddy.

Toddy, which has different meanings in different parts of the world, is a drink obtained from the coconut palm. The coconut tree plays just as important a part in the diet of a European as it does in that of the native. Both drink coconut milk and use the coconut with curry and in salads, while every native—and every wise European—has his own toddy palm from which he obtains his daily quota. Getting toddy from a coconut palm means sapping the life of the palm concerned, so that it never bears fruit. At dawn and dusk the natives climb the palms, knives between their teeth, to collect the full gourd of toddy and replace it with an empty one. The shoot has to be cut regularly twice a day so that the rising sap can be drawn off.

The most reliable book on living in the British West Pacific islands is *It's A Bigger Life,* by Lucille Iremonger. Mrs Iremonger, who, with her husband, spent some time there, paints a fascinating picture of island life, full of humour, but not forgetting the snags hinted at in these pages. In the foreword she says:

> This is a very personal little book. It tells of the everyday lives of two very young people, newly married. In the ordinary way these things would be of no interest to anyone but themselves. But by chance the first year or so of our marriage was passed on lovely Pacific islands—which, incidentally, were shortly to be taken by the Japs... Perhaps, also, the tribulations of an inexperienced young woman of soft and civilised tastes, and in her own eyes utterly miscast for adventure, suddenly landed with the part of Pioneer Woman, may provide some entertainment for a wider public.

The book does much more than that. It should be in the library of every would-be escapist. Mrs Iremonger has told me herself that she went to the South Seas without the illusions most of us are still apt to cling to. 'I knew that the life I should lead there was the sort of life I had least interest in,' she said, 'and in many ways actually loathed. I didn't go out looking for a South Seas honeymoon.'

Her humour must have been a priceless asset during her Pacific sojourn. She lived for more than a year without fresh meat or vegetables, without a doctor, dentist, shop or other woman of her race on tiny coral islands, entirely cut off from the rest of the world. All this would have been an adventure enough for most women, but Mrs Iremonger also learned the native languages, and taught natives algebra in exchange for lessons in fishing and wireless telegraphy. She experienced changes of climate that prostrated her on days of oppressive heat and swamped her out of house and home in devastating storms. Sometimes she and her husband lived for a whole fortnight on nothing but rice and coconuts and an occasional fish. She nursed her husband through scurvy and cured a Chinaman of beri-beri. She herself was seriously ill, dying, so she thought, and her husband had to radio messages to distant outposts to request medical assistance. Fortunately they were on one of the three islands (Vaitupu) of the Gilbert and Ellice group that possessed a portable wireless set. The wireless batteries were charged by means of a boy pedalling a stationary bicycle. The chief point about this recital of a fight for survival is that this married couple were not beachcombers. Mr Iremonger was in fact an administrative officer in the British Colonial Service.

It was fortunate indeed that Mrs Iremonger was such a resourceful and practical woman. Any woman who hankers after a South Seas life should listen to her hints on household problems in this area. She points out that tinned and luxury foods are exceptionally expensive in the Pacific—current prices for foodstuffs, except for basic foods, all tend to be well above European or American prices—and unless you have a more than adequate income, you must, as Mrs Iremonger points out (and it applies even more so today), be prepared to adapt your menus accordingly. That the British culinary genius, so often maligned because it is hidden rather than flaunted, can rise to great heights even with modest materials is proved by the ingenuity of Mrs Iremonger's menus. In her book she mentions sixteen fish, egg and meat dishes, eight ways of making vegetable dishes and sauces, fourteen puddings and seven cakes.

Of all the islands where the British influence still exists I would recommend the Fiji group first of all (see p87). There are about 225 of these reef-girt isles in the South Pacific, their climate is as good as any and they have reasonable amenities. The golliwog hairstyle of their natives is unique. They are a hardy race, less attractive than the Tahitians and certainly less romantic, but they make first-class rugger players, which is a compensation for devotees of the game, and take the game very seriously.

Occasionally there are islands for sale, or to be leased, in Fiji, but rather fewer than was the case some years ago. In early 1972 there were two islands for sale—prices ranging from £1,100 to £75,000 (for a 1,700-acre copra-producing island). Most of these island deals are done on the spot and through intermediaries. It is no use writing to the local administration. You should visit Suva, the capital, or write to the editor of the *Pacific Islands Monthly* (see Appendix II), who could probably put you in touch with agents, or, better still, advertise in the magazine.

On the other hand, if you prefer to settle on an island without attempting the hazards of ownership, Fiji does offer a haven for the civilised man or woman. A naval commander who retired there on pension wrote and told me: 'The cost of living certainly isn't cheap, but I don't know of a better hole. I have lived in a good many places in the world and I still plump for Fiji'. This group of islands has the advantage of being more accessible than most others in the Pacific, for Suva, with its air connections with the USA, Canada, New Zealand and Europe, is the air junction of this vast ocean. It is not such a cultural abyss as most island capitals and possesses clubs for yachtsmen, golfers, rugger enthusiasts and chess players, a scientific group and an arts theatre centre.

Most of the land in Fiji is owned by the tribal chiefs in perpetuity—a present from Queen Victoria—but there are some freehold sites, notably at Deuba, along the coast between Suva and the airport at Nadi. Buyers of residential plots here must build within three years. The selling pitch is aimed both at people who intend to retire and at investors who want to build and then let to visitors. You can buy plots sufficient for building a house and creating a garden for between £3,800 and £8,000. There are mortgage facilities available and Pacific Hotels and Developments Ltd has a London office in Hanover Square. But it should be remembered that newly independent Fiji is anxious to develop her tourist potential and, though this island is never likely to become another Hawaii, it will probably be extensively developed.

A typical freehold island in the Fiji group offered for sale recently was Naitauba, which lies only 15 degrees south of the Equator and about 180 miles north-east of Suva. Some 3,000 acres in extent, Naitauba is planted with coconuts, and, as well as giving a regular income from the sale of copra, the island's stock (part of the sale) included a herd of mixed bullocks and a small Jersey herd. There are a bungalow, houses for labourers, lorry transport, a launch and an electricity plant. The price asked was £70,000 ($168,000), which was not excessive by Fijian standards. As a matter of interest to television-watchers this island was bought by the American actor Raymond Burr, hero of *Perry Mason* and *Ironside*.

One important factor that should not be overlooked when considering the Fiji group is that there is no malaria on the islands, no poisonous snakes and low income tax, no land tax or capital gains tax and no strategic military installations.

One of the principal island-owners in this area is Mrs Phyllis Malley of Hawaii (see Appendix II). She has had as many as four islands on the market at once, ranging from Namuka Island, only 1 mile from the mainland of Fiji and 6 miles by motorboat from Suva, costing £104,165, to Kora

Levu, a 5-acre unspoiled islet, costing £15,000. Namuka, which is large and has fresh water springs, is really a tycoon's island, but Kora Levu is the legendary summer home of Dukaronga, whom the Fijians call the Shark God, and is conveniently situated in the Somo Somo Straits, a mile from Taveuni.

Two other islands belonging to Mrs Malley were advertised at £81,250 (Viubana) and £39,583 (Nawaci). The former is 75 acres in extent, situated in a sheltered position close to the northern tip of Taveuni Island, which you can wade to at low tide. Included in the price was an iron and wooden dwelling-house, with a 4,300-gallon concrete water tank. Nawaci, though smaller (26¼ acres), is said to produce about £500 worth of copra a year, and lies off the southern coast of Vanua Levu. The advertisement described it as possessing 'a Fijian-style dwelling made of reeds and palm leaves, not nearly as fragile as it sounds, and surprisingly well-equipped with shower-room, wash-basin and toilet, built-in beds, wardrobes, tables, etc', with fresh water piped from the mainland.

Worth visiting, if you go to Fiji and wish to explore further afield in the Pacific, are Norfolk Island and the island kingdom of Tonga. In Tonga, the only self-governing kingdom within the Commonwealth, you might be welcomed if you are a technician of some sort. Tonga's isles were discovered by Captain Cook, who, out of gratitude to the inhabitants, named them the Friendly Islands on account of the enthusiastic and hospitable welcome he received here. This is another Polynesian group south of Samoa and south-east of Fiji.

Not many islands advertise the escapist opportunities they possess, but Norfolk Island, which seems as remote as Pitcairn, is recommended in its official guidebook thus:

> To people with small settled incomes, such as those on the retired lists of the Navy, Army and Civil Service, Norfolk Island certainly offers an ideal home, and many such from the United Kingdom and other parts of the Commonwealth already reside in the territory. The cost of living is low.

For years Norfolk (an Australian dependency) was regarded as a tax haven, with no income tax. This issue was raised in the High Court of Australia in a case in May 1971, when the plaintiff challenged the right of the Commissioner of Taxation to make inquiries on Norfolk Island and also to test the validity of an assessment passed against a Norfolk Island company. Companies had mushroomed on the island because of the freedom from taxation, the number of Australian firms registering there having risen from sixty-seven in 1966 to a peak of 400 in 1970. Then in 1971 the Deputy Commissioner of Taxation in New South Wales issued a number of final notices demanding lodgement with him within twenty-eight days. Such returns had not previously been called for from Norfolk Island companies, and the act was challenged in the aforementioned court case. The Australian Government eventually ended Norfolk Island's tax exemption in 1972.

Norfolk Island, which also has close links with the *Bounty* mutineers, has a climate somewhat similar to Madeira, and is in many ways the most temperate of all Pacific outposts. In addition to the main island there are two smaller ones—Nepean, a low islet of coral sandstone, lying about ½ mile south, and Philip Island, a volcanic mass 900ft high, 3-4 miles south. There is a wide variety of scenery in Norfolk island—tropical vegetation in the valleys and, higher up on the plains, the sort of temperate-zone country you would expect to find in Kent. The islanders, all of whom speak English, are a mixture of West Country English and Tahitian. The climate is particularly good for invalids and the island has a first-rate health record. Malaria, hookworm, filaria and other typical Pacific diseases are unknown.

Norfolk has two golf clubs, several tennis and croquet clubs and excellent deep-sea fishing. It has a hospital and a cable station. Some cottages are available for renting at relatively low cost—between £3.50 and £6 a week ($8.40-$14.40).

Anyone who considers living in isolated groups of islands ought not only to be able to handle small craft and understand marine engines but also to know quite a lot about wireless telegraphy. They should be able to transmit messages, understand morse and do simple repair work on radio sets. Your ability to handle a set may make all the difference between life and death one day, and, as a hobby, wireless telegraphy can bring you many friends. For instance, Dr Thomas Davis, medical officer in Raratonga (Cook Islands) and one of the chief amateur radio fans in the Pacific, has established communications over many years with a wide range of radio hams all over Oceania. Wireless telegraphy is also useful as a job in the Pacific, for radio operators, like doctors and dentists, are not easily obtained. If you are an experienced radio operator, you can be fairly sure of getting a job at lonely Pitcairn Island, home of the *Bounty* mutineers. Few of the radio operators who are sent to Pitcairn seem to stay, yet there must be somebody who would welcome this type of life.

You could hardly find anywhere more completely cut off from civilisation than Pitcairn. Its nearest neighbour is Mengareva, 300 miles to the west, with no communication between the two islands, while to the east the nearest inhabited island is Easter, 1,000 miles distant. To the north and south you would have to travel 2,000-3,000 miles before sighting land. Desolate though it may be, Pitcairn has one of the most peaceable and disciplined communities you will find anywhere in the world. The inhabitants are Biblical Essentialists who have lived according to strict Biblical teachings ever since John Adams and his ten Tahitian wives decided that the island had suffered enough through the primeval promptings of man. But it took an era of looting, torture, rape and murder, ending with John Adams as the sole surviving male, before they learned that crime does not pay. It was a lesson that Adams took care to inculcate into his descendants, who have never transgressed since. No liquor or tobacco are permitted in Pitcairn and profanity, scandalmongering and quarrelling are prohibited by law.

The Pitcairn language is a mixture of pidgin English and Tahitian. It is clipped and idiomatic, but it is a polite language. There is not even a word for 'lie'. If you want to accuse someone of telling a lie, you say 'Es stolly', which means, roughly speaking, 'Tell me the old, old story'.

Pitcairn is under the civil government of New Zealand. No land may be sold to an outsider and no land is owned outright, or in fee simple. The land is parcelled out for cultivation by the inhabitants according to the size of their families. The community is practically diseaseless—there is neither insanity, tuberculosis, venereal disease nor leprosy, and even childbirth is reputed to be exceptionally easy and painless. Perhaps nature has decreed rightly, as the island possesses only one competent nurse and neither a doctor nor a hospital. Everything is done communally. For example, the inhabitants all fish on Wednesday, gardens are tended on Thursdays and the weekend food is prepared on Fridays.

Sometimes one or other of the cable companies has an island for disposal. One fairly recent example was in 1964, when the British Colonial Office and Cable & Wireless Ltd offered an idyllic mid-Pacific atoll called Footprint of Heaven for sale. The cable company, rather prosaically, called it Fanning Island, and described its situation as 3,000 miles from any mainland and 170 miles from the nearest radio station, Christmas Island. Footprint of Heaven is at no point more than 10ft above sea level, yet it boasted a substantial modern village with a guesthouse, a huge deep-freeze, swimming pool and floodlit tennis court. The Colonial Office described its sale as 'an extremely rare event', brought about by the opening of the Commonwealth Pacific telephone cable, and the company no longer requiring a booster station on the island, which had been inhabited by some fifty-seven Cable & Wireless personnel. The price asked was £60,000 ($144,000).

NEW ZEALAND FIORD ISLANDS

For out-and-out escapism you could do worse than take a look at some of the islands in the New Zealand fiords—either as a permanent home or as a holiday retreat (see p87). The islands have no escapist tradition and are so remote that even New Zealanders know little about them, but they are ideally situated. If you should visit this part of the world, go to Preservation Inlet, a fiord at the extreme south-west corner of New Zealand, and learn about Mr Jules Berg. For the best part of half a century he owned Preservation Inlet and all the islands in this fiord, which were situated in Fiord County, a vast tract of rugged land covering more than 8,000sq miles and for years inhabited by less than a dozen people.

Mr Berg's fiord was some 20 miles long, a deep narrow arm of the sea winding inland between precipitous slopes clad in dense forest. It has so many islands that a whole colony of hermits could live here without intruding on each other's solitude. For years Mr Berg had the place to himself: the nearest village was 50 miles away as the crow flies and it would have had to fly over some of the toughest rock-jagged and most heavily forested country in the world to get there. It is so little explored that the supposedly extinct bird, the notorni, lived in one of the most inaccessible parts of the region for fifty years after it was thought to have died out without ever being sighted by human beings. The only practical access to Mr Berg's domain was by sea.

Yet he established for himself a comfortable cottage, with an extensive vegetable garden adjacent, and a highly ingenious deer trap that ensured him a constant supply of fresh meat from the surrounding forests. Rare visitors to his hideout invariably left with a profound respect for the qualities of his parsnip wine. Moreover he had a goldmine near his cottage which he worked as often as he felt like it. Surely no escapist was ever better equipped to face adversity than Mr Berg and he was still profiting from this in the early 1950s.

As civilisation produced very little that was of any real value to him, he could afford to let the gold stay where it was most of the time.

The case of Mr Berg proves that there is still scope for the out-and-out escapist. If you like loneliness, towering beetling cliffs and mountains, fiords that are often shrouded in mist and a certain rugged grandeur of Nature, then you have all this in the islands of the New Zealand fiords.

THE ENCHANTED ISLES

In defiance of all the objects Nature throws in the path of the escapist in the Pacific, and possibly because of that obstinacy inherent in human nature, these lovely will-o'-the-wisp islands attract more genuine escapists than any other similar far-flung group in the world. To the Pacific, despite bans, immigration law and climatic and hygienic disabilities, come year after year the would-be builders of Utopia, the seekers after paradise and the re-creators of Eden.

The islands most favoured by this politico-philosophical type of island-seeker are *Las Islas Encantadas,* as the Spaniards who first discovered them in 1535 called what we know today as the Galapagos group. They were given their first name because, owing to faulty navigation on the part of the ancients, they were always 'disappearing', but later the word *Encantadas* came to have a much more romantic meaning: to escapists from Europe it became synonymous with 'enchanting' or 'mystic'. The name, however, does not seem to have misled Darwin, who spent a month in the group, calling them the 'Ash Heap' of the Pacific and stating that they were 'never intended for human habitation'.

There are thirteen islands in the group, which lies 730 miles west of Ecuador. Galapagos was for long the post-office of the Pacific. The old whalers came here and left their letters in a barrel, to be collected by homeward-bound ships. The islands were at one time swarming with enormous tortoises, some of which have found their way to the British Zoological Gardens.

It is odd that so unpromising a group of isolated equatorial isles should have lured so many romantics to their shores. The answer is that here, it was felt, an anarchical life was possible without the interference of civilised busybodies. So, apparently, thought Dr Friederic Ritter, a German dentist and physicist, who had for years planned an ideal community in which he could put into practice his theories of a new way of life, which seem to have been influenced by Bernard Shaw, D. H. Lawrence and Lao-tse. He came to settle on Charles Island in the Galapagos, taking with him as his chief disciple the young wife of a middle-aged teacher, who was a former patient. Ritter's wife agreed to accompany the pair as their housekeeper. It was an odd amoral trio who made this experiment in living, but their regime was almost puritanical in comparison with those of later comers to Charles Island.

Dr Ritter chose Charles Island deliberately because he wanted 'the solitude of an almost desert island in the Far Pacific, the opportunity to cultivate our reflective powers to the fullest, which are denied to man by the complexities of modern life'. The doctor may have been a philosopher, but he was also a practical man. He hacked a site for their home out of the volcanic surface of the island, tapped the water of a spring, developed an orchard and irrigated the land by means of a series of channels fed from a spring. He and his companions were all vegetarians and here they seem to have grown what are not always easy to cultivate in the Pacific islands, fresh vegetables, though greens were only obtained by a substitute plant for cabbage. Having achieved this much, the doctor devoted most of his time to writing, and later he was to publish an account of his life on Charles Island.

News of the Ritter experiment trickled through to the outer world. Other people, with similar ideas, wished to follow. They even wrote to the doctor asking for advice, The newspapers took up the story of the Ritter *ménage*.

Then came the other settlers. First were five young Germans, who occupied the caves on the mountainside near the Naranjal crater. Next arrived a German woman with her sick husband, an Indian woman-servant and a menagerie of monkeys, parrots and dogs. Others followed, but most of them gave up the attempt to make a life there after half-hearted efforts. The blonde seductive self-styled Baroness Eloisa Bosquet von Wagner, however, was made of sterner material. She arrived from Paris with two young Germans and an Ecuadorian manservant, determined to share in Dr Ritter's philosophical experiment. The doctor, angered and not liking the lady in the least, forbade her settling in his 'colony'. She went to another part of the island where she lived with her 'loves' or 'hates'—if one judges by the strange methods by which she sadistically persecuted her companions. Strange stories of beatings and torments filtered through to the other inhabitants of the isle, screams were heard in the night and with them mingled the sounds of a cracking whip. The Baroness called herself 'Queen of the Island', ordered away other newcomers and threatened everyone save the doctor with the pistol she always carried around with her. What happened ultimately is still somewhat of a mystery, but it now seems certain that Lorentz, one of the two Germans, murdered the other German and the 'Baroness' and hid their bodies. Lorentz himself was later wrecked with some Norwegians on Marchena Island, where, without food or water, he perished. Months later his body was found there.

Dr Ritter died on the island after eating—ironically enough for a vegetarian—poisoned chicken. Drought had forced him to break his vegetarian habits and the chicken was diseased. His companion, Dore Strauch, left the island after burying him in the orchard. She told the whole story of this weird island drama in her book, *Satan Came to Eden.*

Yet still the 'Enchanted Isles' enchant some people. In 1926 a contingent of eighty-six Norwegians came to settle here, thinking they would find a paradise of plenty. They brought cows, chickens, farm tools, seed and a tractor. Several died within a few months, others left and died later from the ill-effects of the climate and the diet, and the remainder went back to Norway penniless. In 1937 an American couple started a farm on James Island, but after staying only five months they were ordered off by the Ecuadorian military governor on the grounds that there was not enough water on the island and that 'he could not give them military protection in such an outlandish place'. An odd excuse when one considers that no similar action was taken with regard to the occupants of Charles Island a few years earlier.

Perhaps the roving seafarer is best adapted to life in the Pacific. He can certainly have his fill of islands and move on when he is bored, so escaping from island claustrophobia, a form of mental depression that affects some people in this area in much the same way as the *cafard* in the desert. There is the advantage that your transport will be welcomed by the islanders, as there is ample room for competition around such scattered outposts. *The Pacific Islands Monthly* is full of news items about people—usually a man and wife team—who are moving from island to island in their boats. These are not members of the smart rich yachting fraternity, but people operating anything from a sailing dinghy to a ketch. Just as the society columns of European and American newspapers are filled with the movements of Lady Blank and the Princess of So-and-So, who are due to arrive in Paris or Rome, so the *PIM* announces that 'Betty and Bill' have just left Papeete and expect to visit Suva and Flint Island before returning to Papeete in the autumn.

There are also the 'forgotten islands' of Micronesia, which since World War II have been American Trust territory. These include Truk, the Marshall Islands, Ponape, Palau, the Mariana Isles and Yap. Of all these Ponape is probably the most attractive, notwithstanding the fact that its main export is scrap metal and surplus war weapons. It is on the hurricane belt and the temperature rarely drops below 80°F, and it is

Page 87
(left) *Nabukeru in Fiji;*
(below) *The fiord isles of New Zealand*

Page 88 (above) *Bermuda beach scene;* (below) *Pigeon Island off St Lucia*

not exactly easy to reach (you get there via Honolulu and Yap or Truk and then by means of a somewhat ancient amphibious aircraft). Ponape is 176sq miles in area, has 18,304 inhabitants and has a better public health service than most other territories in the Pacific. Its interior is breathtakingly beautiful, its woods filled with some of the rarest of wild orchids and parakeets of many colours, and rising to the half-way point of hills that are violet-hued in some seasons and bright emerald at others. Its people live in farm communities rather than villages. I am reliably informed that the US Government will encourage anyone wishing to set up pepper-gardens in Ponape.

Finally, if you get the chance, take a look at Catalina Island, a small but quite perfect place lying some miles south-east of San Pedro Bay, California. Most maps do not show it, but it should stand high on anyone's list of islands to be inspected. Its principal resort is appropriately named Avalon, and the island is so exquisite, so completely typical of a dream island, that it is frequently used by Hollywood for tropical background in films.

5 THE WEST INDIES AND THE WESTERN ATLANTIC

One of the first problems that besets the island-seeker is transport. He has found the perfect island, but its remoteness and his own lack of transport make him reluctant to settle there. Probably he hesitates about buying a boat more through lack of knowledge of seamanship than the high cost of maintenance.

In the West Indies and the Western Atlantic, however, an island home without some sort of a boat is not as a general rule a practical proposition. So you had better make up your mind from the start either to find the answer to the transport problem, or else choose an island so close to civilisation that you can row there if necessary. The latter was the solution adopted by Mr Stanhope Joel, the racehorse owner, though I do not say he arrived at this conclusion by the same processes of thought as you and I. Mr Joel decided to leave Britain because of high taxation and so he bought a 7½ acre island named Perots, which is only 100yd offshore from Bermuda. 'It is actually within swimming distance,' said Mr Joel 'and you can cover the course easily in a rowing boat.' There are other tiny isles off Bermuda capable of being turned into homes. Mr Fred Bennett, formerly MP for Torquay, acquired Oswego Island in 1956 as a place for retirement. It is 4 acres in area, covered with pine trees, and has an orange grove, a pink beach and a two-roomed cottage.

Since World War II there has been an increased demand for island homes in this part of the world and prices have

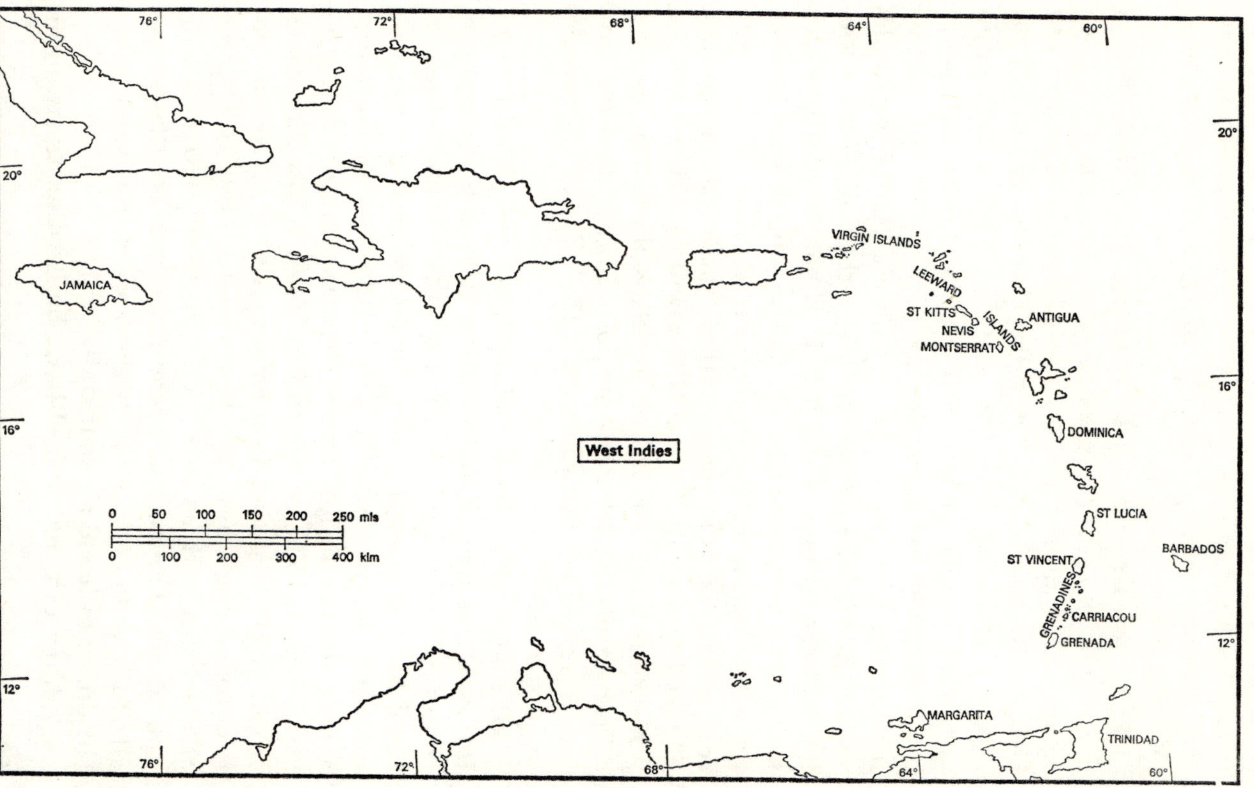
West Indies
76°
72°
68°
64°
60°
20°
16°
12°
JAMAICA
VIRGIN ISLANDS
LEEWARD ISLANDS
ST KITTS
NEVIS
MONTSERRAT
ANTIGUA
DOMINICA
ST LUCIA
BARBADOS
ST VINCENT
GRENADINES
CARRIACOU
GRENADA
MARGARITA
TRINIDAD
0 50 100 150 200 250 mls
0 100 200 300 400 klm

rocketed. Nowhere else in the world at the moment do islands fetch such a high average market price, and the value of land is such that the mere possession of an island is a good investment.

To return, however, to the importance of a boat, most of the islands for sale are situated far from normal shipping routes and transport of your own is essential. Also West Indian waters are a sailor's paradise, and sailing is one of the chief joys of life out here. I use the word 'sailing' with a purpose. Most people are put off by the word 'yachting', which conjures up visions of a highly expensive sport, a millionaire's income and conventional attire with all sorts of petty rules and codes to be observed at the local yacht club. Forget yachting and concentrate on sailing. It is not necessarily a rich man's hobby, as should be obvious to anyone who observes the artisan sailor in the smaller harbours round the south coast of Britain. Many of them run a small craft on less than £1,000 a year total income. Sailing will provide you with a worthwhile recreation and for an island life you need one—sailing, gardening, birdwatching, or fishing. Unless you have such a recreation, then you are not likely to find the peace of mind you seek.

Why not build your own sailing boat? You probably think this is a job that should only be tackled by professional craftsmen, but today it is quite possible for the amateur to build his own boat cheaply. You do not even need to be a carpenter, for you buy the parts of the boat prefabricated, and the shell of the hull comes to you already shaped with the correct curves. The prefab racing *Firefly* is just the answer. It can be put together in a garage. A crate arrives at your door containing a boat, which, at this stage, consists of pieces of wood that have already been planed and bevelled to the right shape. The necessary screwholes have been bored and all you have to do is to follow the book of instructions, then fair up, varnish and rig your craft.

Children can, and indeed have, built sailing boats to this

plan. For the amateur adult the job of putting the parts together, glueing and screwing, rigging and varnishing takes no more than about 80hr. The finished job, with a ready-made mast and rigging for a 12ft sailing boat, can cost a quarter or less than it would if the boat came straight from the salerooms.

If you want to use your boat for fishing in a big way, then you will require something more ambitious and it will be necessary to buy a craft. It is fairly easy to pick up bargains in a variety of all-weather craft, petrol-driven, throughout the West Indies. Some island settlers have made their boats their homes and so saved rent, an idea that may commend itself to the younger escapists. If you can get hold of some of the surplus landing-craft that come on the market from time to time, you will find these are easily convertible into homes and can be made into quite comfortable living-quarters high and dry on a beach owing to their flat bottoms. Equally they are easily navigable in shallow waters and for an amateur, the problem of going aground in a landing-craft presents no major difficulties—you either wriggle off with your engines or wait for the tide. Or, if your kedge anchor has a reasonably strong winch, you can haul yourself off. But—a word of warning here—it is not advisable for an amateur sailor to venture far from land in the vicinity, say, of Bermuda. Gales spring up suddenly and acquire a terrifying force, and, unless you are a first-class astro-navigator, it is the easiest thing in the world to miss the small low-lying mass of Bermuda, the only stretch of land in the area. I once experienced a sudden sharp gale only 10 miles out from that island and was forced in a 60-ton craft to remain hove-to with a sea-anchor rigged for a matter of some 10hr, and during that time we were nearly hit by a waterspout.

BERMUDA

Bermuda has much to recommend it. Despite its relative isolation, it has all the amenities of a civilised life. Even if you cannot buy one of its satellite islands like Perots, it is small enough for you to enjoy a typical island life on a small estate. It is not tropical, but has a mild climate all the year round, the average temperature being 70°F, with a daily range of about 10°F. Snow and frost are unknown and there is no long rainy season. It has had a reputation as a health resort for years and the Bermudians on an average are long-lived. The *New York Medical Journal* states: 'The colony passed a 100 per cent. test as a sanctuary for hay-fever sufferers'.

Yet even in long-peaceful Bermuda (see p88) there have in recent years been discordant notes: the Black Panthers have made efforts to spread their propaganda there and it could well be that some future regime will not tolerate what has for so long been a white-dominated island. It may well have to adapt itself quite considerably in the light of its independence, and its eighteenth-century atmosphere, with its graceful and leisurely tempo of living, may be an early casualty.

Sporting attractions in Bermuda are sufficient to satisfy most tastes. There is excellent sailing, the Royal Bermuda Yacht Club being a pleasant rendezvous. The Tennis Stadium has attracted some of the world's stars, and the island has a game-fishing tournament that lasts most of the year. Bermuda is the summit of a volcanic mountain range, and you can play golf on the tip of a subterranean volcano. The Bermuda Golf Course is in this respect unique and the pleasure of playing is heightened by breezes that carry the scent of tropical flowers. Only Charles Blair MacDonald could have planned the Mid-Ocean golf course and only he—with the backing of the Furness Withy Steamship Company—could have achieved it. The sea near the course is fringed by delicately tinted pink and white sands, which trail back into clusters of brilliant wildflowers and gracefully swaying palms and shady cedars. There are magnificent challenging greens that call for every type of approach shot. MacDonald found his task of creating a course here by no means easy. Below the 6in of soil was coral rock, but when the job was done he was able to write: 'I doubt if there is an eighteen-hole golf course which will equal, certainly not surpass, from a golfer's standpoint, the links, in any semi-tropical clime, nor at any health resort'.

The late Hervey Allen, author of *Anthony Adverse,* was a frequent visitor to Bermuda. His honeymoon lasted so long that a New York newspaper reported that 'Mr and Mrs Harvey Allen have just returned from their Bermuda honeymoon with two children and several dogs!' It had been five years' blissful start to married life. Perhaps the tradition, sentiment, whimsy and lyrical quality that mark Bermuda house names convey something of the spirit of the island. Hervey Allen called his house Felicity Hall. There is Queen of the East in East Broadway, and the nautical touch is provided in such names as Briny, Blue Water, Spindrift, Sea Spray and White Caps. Landfall, too, is a happy and euphonious name for a home, as is naval nomenclature like Mizzentop, The Binnacle, Shipshape, Sundeck and Hove-to. One house was named Thirty-nine Steps by a John Buchan fan, though the number of steps has now increased to forty-one.

Apart from the possibility of buying small islands like Perots—within inflatable-dinghy range of the main island with the help of an outboard motor—there is a large selection of modern well built houses for sale and to rent in Bermuda. But land prices are high.

THE EASTERN CARIBBEAN

In recent years there has been a boom in islands in the Eastern Caribbean, but gone are the days when you could buy an island of 15-20 acres off Grenada or St Vincent for between £200 and £300, or a bargain like Lord Baldwin's when he bought Dead Man's Chest for £100. Generally speaking, the price of islands has increased by six times in the past twenty years. Just one example. A British naval commander some years ago bought a 12 acre island off

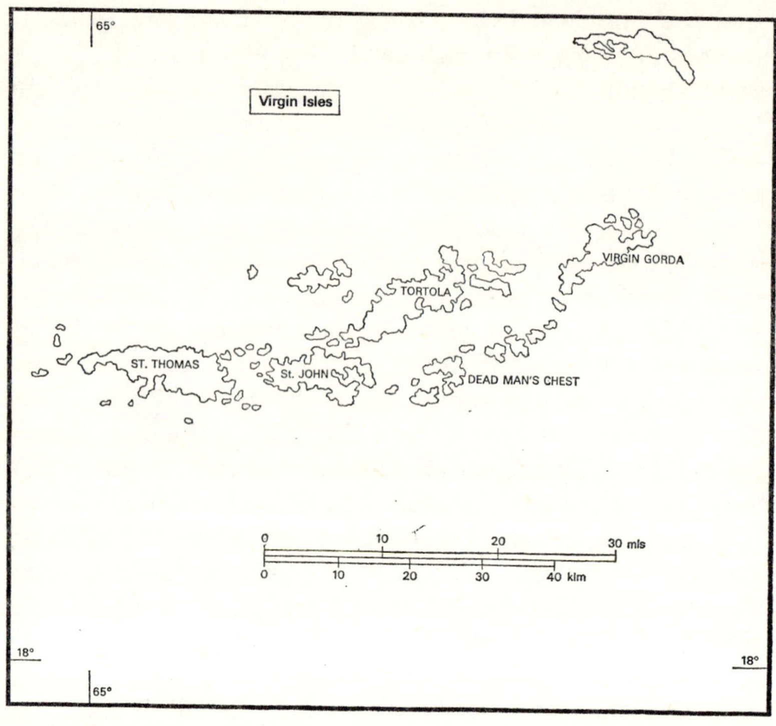

Grenada for £12,500 ($30,000): the very same island had been sold by another naval commander for £2,000 ten years previously. Many islands off Grenada, West Indian home of nutmeg, spices and cocoa, and St Vincent, famous for its arrowroot and sea-island cotton, are privately owned. Those near St Vincent belong mainly to local families and a number of those near Grenada belong to cocoa planters or business-men in St George's, the capital.

Such islands do not always sell quickly—one quite small island sold for £12,500, but only after its owner had spent £1,500 on advertising. The islands off Grenada have the advantage that they can easily be reached from the main island by a small boat with an outboard motor. Some have their own private beaches and are covered with coconut trees, nutmeg and ground crops. Others are overrun with grass and shrubs and rise sheer from the sea. Those owned by businessmen are used for weekends or holidays. There is good sea-fishing here and ashore there are plenty of wild pigeons to shoot.

On the whole this part of the Caribbean offers some of the best bargains in islands, which are much less expensive than those of the Bahamas and elsewhere. They have the advantages of being near civilisation—the South American mainland—and of having a tropical but reasonably pleasant climate. On some of them you could supplement your income by growing crops of cocoa, mace and nutmeg, and many of them will support cattle. From the air the islands off Grenada and the Grenadines look like Capri and the small islands in the Sorrento Peninsula. Most of them are deserted, but here and there a stone house with cultivated grounds, shimmering sandy beaches and gaily flowering trees calls the traveller with all the enticement that the Sirens used when they sang to Ulysses.

In shape the islands vary: some rise from the blue Atlantic like the single hump of a camel, while others look like a tiger skin lying with paws outstretched before the fire in an English manorial hall. Glover Island, only a few miles from St George's, was used half a century ago as a whaling station. Every winter whales came down in herds for a winter cruise to the Grenadines and the industry might still be thriving if the whalers had not thrown the bodies of the whales back into the sea after the insides had been removed. In spite of the large number of sharks that frequented Glover Island as a result, other whales coming down from the north would see their dead relatives and change course. Today Glover is privately owned by an Englishwoman.

There are some thirty Grenadines belonging to the islands of Grenada and St Vincent. One delightful and altogether charming dependency of Grenada is Carriacou, which is as far from Grenada as Deal is from Boulogne. It is 9,000 acres in extent and mountainous, and produces sea-island cotton, from which men's most expensive shirts are made, and limes. Quite often an island with lime-farm thrown in is advertised for sale. Petit Marrunique, a few miles to the north of Carriacou, is an excellent fishing station, and several fishermen-landowners live high up in its hills.

Diamond Rock, which rises 670ft out of the sea, has been given another name by sailors: on a rough day local skippers will say it is really 'bad' close in to 'Kick-'em-Jenny', and give it a wide berth. Not far from this rock, named after the local lady mules, lies Jinny, a bird sanctuary inhabited by pelicans, boobies, cormorants and man-of-war birds. There are scores of these rock islands—The Steer-Clears, The Sisters, The Aunts, Round Isles and Isle of the Quails. There is also the oddly named London Bridge, which is less than 5 miles from the airport and has a hole right through it, caused by hundreds of years of wave erosion.

To fly over the islands of Grenada just after sunrise is to glimpse paradise, and to travel by boat round the islands that dot the 60 miles of sea between Grenada and St Vincent is one holiday the hunter and fisherman ought not to miss. All this is island-hunting territory.

The whole of the Eastern Caribbean group—the Virgins, St Christopher and Nevis, Montserrat, Dominica, St Lucia (see p88), St Vincent, the Grenadines right down to the islands of Margarita (Venezuelan) and Trinidad off the delta of the Orinoco—offers chances of cheaper living than the more northerly islands of the West Indies. As you go south, too, the formalities begin to disappear, and with them go the socialites and the slaves to planned living. Oddly enough the loudest antagonist of a planned economy and social structure is usually the chief protagonist of a rigid code of life. He insists that he shall be free from controls only to inflict them on himself and others around him with all the tedious claptrap of dressing for dinner, the code of calling and leaving cards (now almost unheard of in the Western World), of insisting that bridge is the passport to acceptance by society, of the well worn rebuke at cocktail parties to those who are trying to escape for a few weeks from the social round—'Aren't you a stranger? Where have you been hiding yourself?'—the implication being that you have no right to absent yourself from his circle 'without leave'. Such people ruin the escapist life.

Not far from Antigua, in the Leeward Islands, is St Christopher (or St Kitts) Island, where there are excellent chances for immigrants with a reasonable amount of cash. I am told that for every American who likes the more formal life of Bermuda and the Bahamas there are ten who are looking for something more truly escapist. For these the attractions of St Christopher and Nevis make greater appeal. Estates can be bought on these islands for £3,500-£5,000 ($8,400-$12,000), and then one can live on them for about £1,500 ($3,600) a year.

St Kitts was the first island in the West Indies to be systematically colonised by the British. In Nevis, in 1787, Nelson, then only twenty-five and acting Commander-in-Chief of the Leeward Islands Naval Station, married the young widow Nisbet. Here, too, was born Alexander Hamilton, son of a Scots planter and chief drafter of the American constitution.

The islands off the Latin American mainland from Guyana round to Venezuela should also be considered. Some of them are being sold for as little as £1,000, and one of 8 acres went for £700 ($1,680). When *Islands for Sale* was published I received a charming letter from Mrs Rosamund Wright, who wrote to say that she was anxious to sell her own Guiana Island, together with a dozen small islets off the coast of Antigua. She had lived there happily for many years, but had decided to return to Britain. Her letter was dated 27 December 1950, but her description of the main island may still serve a useful purpose:

Guiana Island is about two and a half miles long by about a mile wide. It consists of several hundred acres. It has a road down the centre. The Great House, which is built of stone, is in the centre. It is several hundred years old, but has been modernised. There is a good water supply, electric light plant, radio, several boats (with outboard motors), two telephones to the mainland. The island is only separated from the mainland by a deep strip of water seventy yards across. The house has four double bedrooms, two bathrooms, kitchen, pantry, big hurricane room below. There is a very large, long sitting-room divided in three by pillars, with a dining-room one end, lounge in the middle and library at the other end. There is a gallery on both sides and wonderful views in every direction.

There is a very fine oil-burning refrigerator, a root-puller (for uprooting unwanted trees). The house is splendidly furnished with some fine old Colonial furniture, good linen and china. There is a Delco Electric light plant, four wood houses, used for servants and extra guests.

There is also a good cow (about to calve), a small herd of deer, a few donkeys and a small flock of sheep (some of them have bred black and white Persians). I have left good servants behind.

There are a few coconut trees and many years ago sugar was grown here. Cotton could be grown. I cultivated an acre of vegetable garden and kept all kinds of poultry here—guinea-fowl, turkeys, chicken and ducks.

I pay a land tax of about £25 a year to the Government for Guiana Island and Great Bird Island, the next largest island. Great Bird is the nesting place of the sea-birds of these waters. It would be possible to live there. It has lovely lagoons for bathing.

It might be possible to live on three or four of the other small islands and keep chicken and goats on them.

The deer on Guiana Island are Virginian deer. Formerly there was a pontoon from Guiana to the mainland on which a car used to cross and drive up the road in the middle of the island. During the war the pontoon rotted and I did not bother to replace it.

Fishing is by net or line or fishpots. There are plenty of lobsters and crabs caught at night with flares.

Pawpaws grow very well and supplied me with plenty of fruit. There is rather a lot of lime in the soil so citrons do not do well. But I have a few lime trees for the limes needed for the house. I also planted a few banana-trees near the house.

I found the climate perfect and there were cool breezes all the year round. I am asking £15,000 for the place as it stands, ready to walk into.

Guiana Island seemed an absolute bargain at £15,000 ($36,000) in 1950, and such bargains can still be picked up in the Eastern Caribbean. There was only one ominous note: the hurricane room! Well, at least protection from hurricanes was provided. Not all island homes have them.

THE BAHAMAS

'Bahamas Real Estate . . . Islands for sale . . . write for booklet, Harold G. Christie, 309 Bay Street, Nassau, Bahamas.' That is just one of the current advertisements, not only in Nassau newspapers but in American and European periodicals as well, all offering islands for sale in the Bahamas (see p105), which must surely be the most prolific hunting-ground for island-seekers from all over the world. From the advertisements it certainly seems as though the whole of this area is teeming with islands for sale and to rent, with firms offering

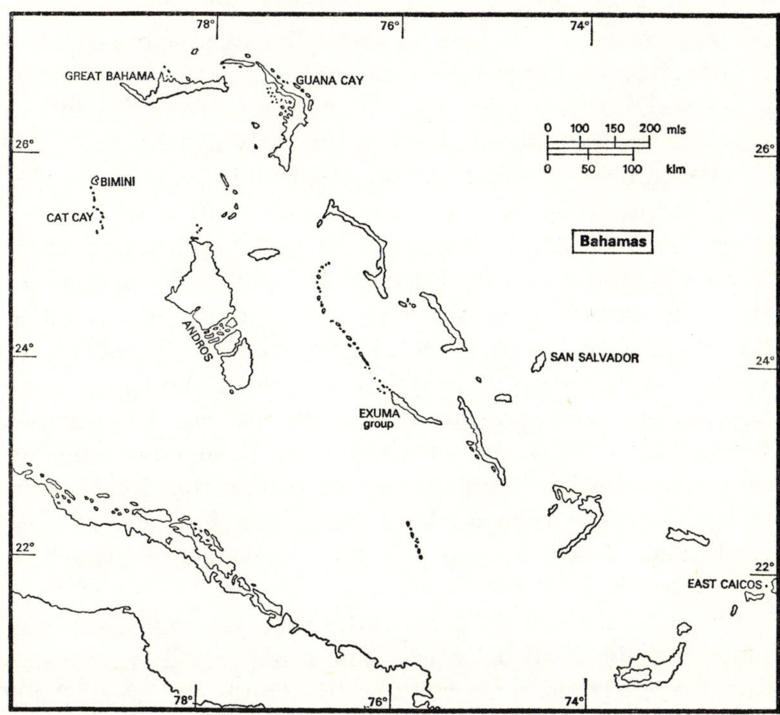

to sell you land with private bays and beaches and countless organisations offering to build you a home.

The Bahamas are delightful but expensive. In the Pacific, as we have seen, you must expect to rough it. In the Bahamas life is made easy: there is no need to work (indeed, you are practically discouraged from working by the immigration laws) and luxuries of every kind are yours—at a price. The importance of earning foreign currency since World War II has made this group of islands off the Florida coast extremely dollar-conscious. Everything has been done to encourage American tourists, and it goes without saying that to compete with them holders of other currencies need to be very well endowed financially. The cost of living is extremely high, though less than in Florida, and you either live luxuriously or stay away. The development of the islands for tourism has been on a considerable scale and much British capital has been invested for this purpose. Great Bahama Island, 60 miles off the American coast, 83 miles long and never more than 4 miles wide, is the scene of the most intense development; so it is best to concentrate instead on the outlying islands.

The climate in the Bahamas is ideal in winter, warmer than Bermuda and its vegetation more profuse and exotic. The season lasts from December to March, the temperature in these months varying between 71° and 74°F. In summer the temperature never rises above 92° and averages 74-80°F. An attempt has been made to create an all-the-year-round season, and this has been fairly successful. You can bathe pleasurably in the sea throughout the year and the summer breezes blow away the sultriness that is so enervating on some West Indian islands. The area is free from insect pests, except for some islands where the sandfly has not yet been eradicated. There are no poisonous snakes, and malaria is unknown.

Of the total 2,800,160 acres in the Bahamas some $2\frac{1}{2}$ million belong to the British Crown. The group comprises twenty-nine large islands, some 690 smaller islands and about 2,400 cays or islets. Only twenty of the islands are permanently inhabited. Some British and Americans own islands here, many using them as holiday resorts. Prospective buyers should make quite sure that the island they buy or rent is free of the aforementioned sandfly, whose bite does not, however, cause sandfly fever, as is commonly supposed. It is quite different from the fever-bearing Mediterranean sandfly.

The annual rental in the Bahamas is determined by the nationality of the lessee and the use to which the land is put. The British, theorectically at least, get preferential treatment, though this is less obvious now than it was twenty years ago. The lowest rental is charged to natives for grazing or stock-raising. There are no land taxes on Crown lands, but there is a comparatively low real property tax based on the assessed value of any building in New Provident Island.

Land is available for purchase, developed or undeveloped, and at present every encouragement is given to purchasers near or distant from Nassau. There is, of course, a good deal of what is called 'buying blind', common in island deals, and usually involving small cheap islands, which can be made to sound attractive in advertisements. People have been known to part with £2,000 on the strength of no more than a colour slide of an island. In the early 1960s the Bahamas experienced a tremendous boom in island and land sales, much of it effected through advertisements in European newspapers, and most of it blind dealing. Islands, cays and mere plots of land were sold to 'blind buyers' all over the world. Many of these people hoped one day to build on their plots or islands, but later found that they could never afford the air fares out to the Bahamas or the high building costs. Others discovered that what they had bought was either miles from anywhere or in the middle of a swamp. This 'blind buying' gave the whole sales operation a bad name and the Bahamas Government issued warnings to people to see what they were buying, or to have the land surveyed by an expert before signing any contract.

There are, of course, many reputable estate agents who can offer good, though not cheap, purchases in islands. Hampton & Sons in London, and Bahamas Properties Ltd and Previews International, both of which have London offices, can be recommended for inquiries. Island prices can range from £156,000 ($374,000) for a 75 acre island like Guana Cay in Elizabeth Harbour in the Exuma group (see p105) to £3,500 ($8,400) for a 1 acre cay. The former included four furnished houses, electric generators and large water-storage tanks.

Other typical Bahaman islands sold in recent years are Little Hall's Pond Island, Ridgeback Cay and Cluff Cay. The first named, comprising 65 acres, is a naturalist's haven, with its wild life, splendid beaches, a lake and good anchorages. It was for sale at £50,000 ($120,000). Ridgeback Cay was only £12,500 ($30,000) for 25 acres but it also had excellent anchorages and good building sites. Probably the most promising island was Cluff Cay, whose 120 acres, close to George Town in the Exumas, with lush vegetation, was for £37,083 ($89,000). There was one ominous note in the advertisement: there were merely 'indications of fresh water'.

These delightful islands are obviously suitable for the residence of those who are, if not rich, at least in possession of an adequate private income. There are very few opportunities for employment in the Bahamas, except for professional men. But if you are rich, do not be put off if an estate agent offers you an island for £1 million ($2,400,000). One such island was Norman's Cay, 905 acres, which has its own marina, clubhouse and airstrip. It was bought by World Land & Investing Co, Ltd (a Bahaman company) in July 1922. The company has started development and, since May 1970, has been offering plots to the public in Britain through Jason Investment Services of Edinburgh.

If you want something really good, you must expect to pay between £12,000 and £70,000 ($28,800-$168,000), but for a small property, needing undergrowth clearance and spring water, there are prospects for £4,000-£12,000 ($9,600-$28,800).

Page 105 (above) *Bahamas coastline;* (below) *Aerial view of George Town on Great Exuma*

Page 106
A tiny bay in Capri

Three small cays, each with building sites about 15ft above sea level and sandy beaches situated within the Exumas Land-and-Sea Park, were offered for £12,500 ($30,000) the lot, or £4,166 ($9,998) for the smallest cay and £6,250 ($15,000) for the other two. Three islands of about an acre each, not far from Abaco, only a 10-minute boat ride from Marsh Harbour and close to Hope Town, Elbow Cay, and connected at low tide, were offered for sale at £6,000 ($14,400) the three. Their proximity to civilisation made these a fair bargain.

Tourism is undoubtedly the best industry for investors in the Bahamas. Some people have put their money into hotels and residential clubs with remarkable success. There is no restriction on the transfer of sterling funds to Nassau from the United Kingdom, or anywhere else in the sterling area. The Bahamas Government issues local currency based on sterling and American and Canadian dollars are generally accepted.

The thorough-going escapist will not like the major tourist areas of the Bahamas, and should give Grand Bahama a wide berth, unless he wants to escape to the fleshpots occasionally. You need, of course, never lack for company or amusement. There are first-class hotels and country clubs, nightclubs and dancing. Midnight beach parties are the fashion. There are bridge and canasta. You can eat all the delicacies of Europe and America, as well as the succulent and intriguing native dishes—Caribbean cooking is a special delight. For sport you have a choice of deep-sea and reef fishing, sailing, swimming, polo, tennis, golf, horse-racing, horse-riding, water-skiing and shooting. Bona-fide visitors may remain in the Bahamas for eight months and may then apply for an extension. Application to become a legal resident may be made after six months' temporary residence, and then you will have to show that you have the means to support yourself.

Largest of the Bahaman islands is Andros, which is 104 miles from north to south and 40 miles wide. Many parts of it have not yet been surveyed, despite the fact that it has a population of between 7,000 and 10,000, and there are thousands of acres of forest. Few white people live here and the chief industry is sponge fishing. San Salvador, or Watling's Island, is where Columbus made his first landfall when discovering the New World in 1492. It lies to the extreme east of the group. Beloved of big-game fishermen are the islands of Bimini and Cat Cay. Bimini inspired the poet Heinrich Heine to write:

> Who's with me for Bimini?
> Step in, gentlemen and ladies!
> Wind and weather serving safely,
> We shall sail for Bimini.

Heine was attracted by the famed Fountain of Youth on Bimini.

> In the Isle of Bimini
> Blooms the everlasting springtime and again
> In the Isle of Bimini
> Springs the all-delightful fountain.

The Fountain is very little more than a legend now, though its healing properties once enjoyed a considerable reputation. Bimini is a fisherman's delight. In the sea around here are to be found the blue marlin swordfish, the giant tuna, the white marlin, wahoo, mackerel, kingfish and grouper. A marine laboratory has been established at Bimini to give scientists an opportunity of studying marine biology at first hand.

The story of Cat Cay Island, lying about 50 miles off the Florida coast, is a romantic adventure in island homemaking. Some years before World War I Mr L. R. Wasey, an American businessman, bought Cat Cay and decided to make it his permanent home. He turned it into a private fishing and holiday resort, with swimming pools, golf course and a luxury clubhouse. So fond of the British was Mr Wasey that he decided to employ all British workmen on the island, which had a happy result for Frank Stokes, barman of the Shakespeare Hotel at Stratford-upon-Avon, who had dreamed of life on a tropical island for years. Mr Wasey, on visits to Stratford, had heard all about Frank Stokes' daydream, and invited him to work as a barman on Cat Cay. Without any hesitation Frank agreed. The bar was run on British lines, and even the village on Cat Cay was based on a typical English model.

A few years before World War I the uninhabited island of East Caicos was leased to a party of Californians, who planned to establish a model community. The colonists were selected from a long list of applicants with qualifications as builders and agriculturalists.

You may be lucky enough to find a small island where a little reconstruction would turn a ruined colonial mansion of a more spacious era into an attractive modern home, but such chances are getting fewer. With the purchase of a Bahamas island goes the prerogative of changing its name. Examples are Miss Bullitt's Goat Cay and Captain Mingo Rolle's God-Rest-Dead-Cay, the latter name coming from the Bahaman phrase 'God rest the dead'.

If you are a treasure-hunter—and few of us are not at heart—the Bahamas in particular, and the West Indies in general, present chances of success that are certainly no more limited than those in a sweepstake. Mere chance has given many Bahaman settlers a surprising and unexpected hoard of treasure, and people still arrive clutching some chart or with some story of treasure handed down from an earlier generation. Much plunder must have been buried throughout the West Indies in the days of the pirates and small amounts are dug up periodically. A poor Abaco farmer became rich overnight after he had secretly sold a small quantity of Spanish gold coins he had dug up in his garden. American yachtsmen found several thousand dollars in the long submerged rusty old safe of a Confederate blockade-runner, located in 30ft of water by a bootlegging plane.

Many Spanish treasure ships were known to have been wrecked in the Bahamas and, in these days of radar, scientific aids to detection and sophisticated diving techniques, it is likely that more treasure will be recovered. Kip Wagner found an old chart of East Florida, dated 1744, showing where the flagship of the Plate Fleet had been sunk, and in the early 1960s led a team that probed the sands and shallows of Florida's east coast. From the remains of the Spanish Plate Fleet that sank in a hurricane in the area he and his divers recovered riches worth more than a $1 million—gold ingots, gleaming doubloons, pieces of eight and even delicate Chinese porcelain that had miraculously survived.

In former years the British Government retained the right to one-half of all treasure found in these islands, but this stipulation has been removed, though finders of treasure are still required by law to report and register their finds. The British Treasury, however, says that the position of treasure-finders is vaguely defined and that there is no firm law covering their rights.

You will find no finer beaches anywhere in the world than in the West Indies, not even in the Pacific. On most of them you will find the answer to the poser about the scarlet flamingo. Usually we see the bird in captivity looking forlorn in feathers of pinkish-grey. Captivity seems not only to dull the spirit of this bird, but to deprive it of its flaming red-feathered cloak. In the West Indies, and particularly on Andros Island, however, you can see these birds in all their scarlet glory of freedom. The 'fillymingos', the natives call them. Whether or not you are an ornithologist or birdlover pure and simple, the scarlet flamingos will not fail to fascinate you with their grace, their colour, their aloof beauty—they make all the difference to the island scene.

Many of the larger islands of the West Indies have been omitted from this section, since in many instances they are self-governing states rather than island outposts. For example, Trinidad is so large and bustling, so teeming with life, that it is more like one vast city than an island. Barbados, perhaps the most English in style of all the islands, is well worth exploring, yet still not quite what most of us are looking for. Jamaica, again, has latterly become a kind of residential area for the cautious escapist who still hankers after civilisation, but somehow, extravagantly beautiful as it is, Jamaica seems too large, too populous and too fraught with acute political problems for our purposes.

In conclusion, take note of the Caymans, where living is fairly cheap and life free and easy, and the islands off Curacoa and British Honduras. They are off the beaten track, they are rarely advertised, and you might have to spend a long time looking for them, but, if you are adventurous enough, they might well suit you. Their prices are often only a third of what would be asked in the Bahamas, and less than in the Eastern Caribbean.

6 THE MEDITERRANEAN

THE Mediterranean offers a unique type of island life. If you have once visited its islands and fallen for their peculiar charm, it is unlikely that any others will give you the same amount of satisfaction. While the Pacific has its romantic protagonists and the West Indies its more sophisticated admirers, the Mediterranean is the natural home for those who want to become absorbed in the way of life of the islands of their choice. You do not go to the Mediterranean to live your normal type of life in new surroundings, or to try out some new theory of living, but because it is nice to do as the islanders do.

You can romanticise the Polynesians, but you cannot submerge yourself in their way of life. You can like Calypso music, adore the West Indies and even become interested in their bizarre cults, but you are still largely the victim of the way of life you came from. You may deceive yourself that you have become a bohemian white man who does not dress for dinner and keeps away from the social round, but you can never become an islander. The respective cultures are poles apart, but it does not matter because you do not need to be integrated.

In the Mediterranean it is different. There are no racial problems, there are no petty snobberies, and social circles are not vitiated by cliques and inhibitions—except perhaps in the jet-set island of Sardinia, which for this very reason will be omitted from our survey. You either become 'one of the islanders', or you return home. This applies to Malta even more than any other island, despite the British association

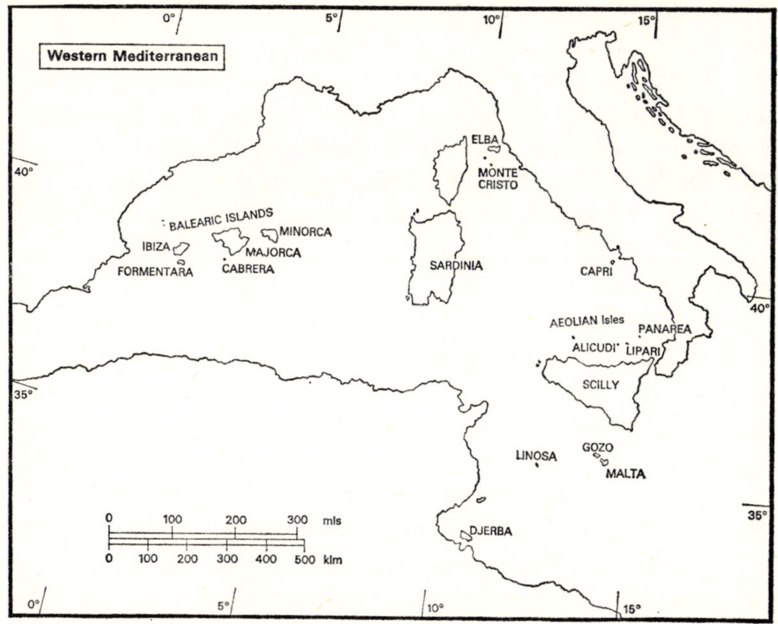

and the fact that English is spoken there.

Therefore the first essential is that you must like the inhabitants as much as the Mediterranean island of your choice. You must enjoy what I should like to call the *camaraderie* of good ordinary wine. Not vintage stuff, but the *vin du pays*, bought by the litre. The wine of the Mediterranean in the islands, perhaps even more than on the mainland, is the symbol of a happy simple life. This is not to advocate a policy of tippling, but a suggestion that you shun hotel cocktail bars with their dry martinis and gins and tonics in favour of some peasant's *auberge*, or fisherman's *taverna*, or sailor's quayside *bistro*, where you may share the *vin du pays* with the friendly folk around you. The *taverna* should become your 'local', the place where you meet your friends and make them.

Wine knows no barriers; it brings people of all classes together as no other drink ever does. It is also the reason for domestic happiness in the Mediterranean: wine inevitably means better cooking, because you use it in the kitchen and because good wine demands good food to go with it. You do not gulp wine as you would beer, standing up in an enclosed taproom, with one eye on the clock. In the Mediterranean they believe that you enjoy wine best in the open air, preferably with a good view and where the sun can make it sparkle in the glass. You sit in the sun, sip your wine and feast your eyes on the sea—at any hour of day or night.

You may doubt whether the appeal of a Mediterranean life is as simple as that. I would simply say that 90 per cent of the people who like wine will like the Mediterranean, while 90 per cent of those who do not will not like it. For where wine is made it is a fact that life is cheaper and that the cost of living is less than elsewhere and this applies especially to the Mediterranean. With the exception of a few expensive tourist isles life here is as cheap as anywhere in the world. For those who are not rich, even for those who consider themselves relatively poor, I recommend Mediterranean islands. All you need for a happy life is a little money, a liking for wine and a love of the islanders and their *patois*. If you can fish, row, swim, keep chickens and are prepared to grow vegetables, even to tend a vineyard, and are easily satisfied with such simple pleasures as a vine harvest festival or a village wedding party, so much the better.

CAPRI

It would be impossible to write about the Mediterranean islands without mentioning Capri (see p106). This lovely isle off the Gulf of Salerno is, despite all the publicity it has had and its steady stream of tourists, still unspoiled and its inhabitants unspoilable. For this reason alone it is worth including in our list. It is, perhaps, the one island whose inhabitants are escapists at heart, escapist in the sense that they do not wish to escape from their paradise, knowing that Capri is worth all the rest of the world put together. They are not escapists who have not realised the meaning or the need for escapism. They are unmoved by political squabbles or wars, consider themselves Capriotes before Italians and are intensely proud of their island.

There are some islands off the western Italian coast for rent or sale, but Capri in itself can give you a satisfaction richer than the mere feeling of ownership. The Capriotes ask for little other than good manners and a certain sensitivity in those who come to live there from the outside world. Living is cheaper than on the mainland and in parts of the island you can feel as remote from civilisation as if you were in far-off Pitcairn. To the historian and archaeologist it is an island of delight; it is a treasure house of history, and not to know the past is not to live a full life.

You must steep yourself in Roman history, in the legends of Tiberius, that man of dark evil passions whose name still evokes a sense of horror among the inhabitants of Capri. You must also learn about the legends of the local wine, some of which is known as *Lacrimae Tiberio,* just as the Vesuvian wine is called the Tears of Christ—*Lacrimae Christi.* They say of a certain vintage of Capri wine that it was haunted by the ghost of Tiberius, which lurks in the caverns deep down below the island. The year it was first drunk in the island men were said to have gone mad with violent remorse for their sins, and wept incessantly. For these wine-bibbers they had a name—'Tiberius-struck'.

You should read Dr Axel Munthe's *The Story of San Michele* for a vivid word-picture of Capri. The guides of Capri are well versed in the history of their island, its folklore and legends. They will show you the 800ft Leap of Tiberius, from which, if the biased versions of Tacitus and Suetonius are to be believed, the Emperor flung his victims into the sea. They will show you all the ancient relics, which are innumerable, and also recount many legends that suggest a fertile imagination rather than a knowledge of classical history. It takes a lifetime of research to uncover the truth behind many of the historical theories about Capri, though it is undoubtedly pleasant to think that Anacapri in the interior of the island was founded by Eros, as the guides will tell you with an air of romantic nostalgia. When you have drunk the wine of Tiberius' tears, you are quite prepared to believe that the God of Love founded this Daphnis and Chloe village exclusively for honeymoon couples.

Capri itself is a curiously shaped island that has been compared to 'a crouching lion watching the Bay of Naples', a sphinx, a crocodile, a Roman Emperor, a dog and a boot. The word Capri derives from the Greek *kapros,* meaning a wild boar. It is 4½ miles long and 2 miles wide, but its smallness is deceptive, for in no other island of its size can you lose yourself quite so easily. Its precipitous cliffs rise to close upon 1,000ft above the Mediterranean, its hills are covered with pines and olive trees and there are vineyards and orchards in its valleys.

The climate is delightful. Never too hot, Capri is the perfect retreat after experiencing the biting heat of the Italian plains and cities in summertime. Breezes are always ready to cool your brow, with one exception—the sirocco. In *South Wind* Norman Douglas describes this pestilential wind, which creates sultry cloudy weather and casts an aura of almost electric mysticism over the island. The wind can produce irritation, but it is mystical in the sense that Daudet found the *mistral* mystical. There is really no other word for it. Hidden away in the depths of cloud and in the force of the wind itself is some subhuman force. The islanders would doubtless blame the maligned Tiberius, but it is more sinister than the most sinister of dead Roman emperors. It has something in it of the spirit of Baudelaire's *Fleurs du Mal.* The Capriotes will tell you that it turns mild men into passionate lovers.

This may well be true, and to understand Capri you must understand its inhabitants and especially its womenfolk. They are entirely without the racial consciousness of most Mediterranean peoples, accepting quite naturally their residents from the outside world. They regard love rather as you and I regard a sudden thunderstorm. I think the Capriote women invented the sirocco legend to ensnare men by auto-suggestion. That may be frivolous, but wait until you experience a sirocco in Capri.

The women have a charm all their own and many of them are exceptionally beautiful—and disconcertingly ambitious. They have a positive talent for rising above their poverty and becoming *grandes dames*. An Englishman who has lived for years in Capri told me:

> They are proud and have an aristocracy of their own. You can find it among women of the poorest homes. Chameleon-like, they are extremely adaptable and quickly pick up the manners and graces of a different mode of living. I have seen many cultured and financially eligible bachelors come to Capri and marry peasant girls. The marriages always seem to be successful and a few years later you will meet a comely lady of the villa on the hillside who is as at home in the drawing-room as in the kitchen.

There has been from time immemorial a small British colony on Capri. Norman Douglas is the most brilliant recorder of the Capriote scene and Gracie Fields has captivated the hearts of the peasants in the same way that many years ago she entranced the Lancashire millgirls and West End audiences. Compton MacKenzie stayed for a short spell and he writes in eloquent terms of Capri. 'The island rises out of the sea with a sublimity that Samothrace alone, of the islands I have seen, can equal,' he declared, and again, speaking of how World War I had affected Capri, he added, 'she had the air of a forsaken beauty, puzzled to know why she was being treated so... It pains me to declare that the charm of Capri is incommunicable. The landscape of Capri is really an inspired chromolithograph. Pen or brush that seeks to render its limestone and blue water is merely gilding the lily.'

THE BALEARICS

The very name Balearics seems to conjure up a gay little tune rather like the start to those champagne bubble inspirations Chopin must have dashed off in his more carefree moments in Majorca—which, though the best known of the Balearics, I do not propose to waste many words on. Palma is like Sandown, except for the heat. Both places cater for the fish-and-chip trade and to go to the Balearics to eat English fish-and-chips seems an unnecessarily expensive outing. Life is certainly cheap if you want to live in Majorca and the winter season is preferable to summer and the tourist invasion, but I would not recommend it for diehard escapists.

Alas, too, Ibiza in high summer is now much the same, ruined by concrete hotels that pop up like mushrooms each year, and a life geared almost wholly to tourism. But of the two islands Ibiza is my favourite, possibly because I knew it before the tourist boom. In its remoter parts you can still escape from the summer invasion and even discover something of the leisurely ancient lovemaking technique of the island belles, which is uniquely their own. Where the Spanish woman flirts behind her fan and uses the language of this graceful shield to convey approval or disapproval, to raise hopes or quell them, the girls of Ibiza, where Moorish blood mingles with Spanish, use their handkerchiefs, peeping out from behind them with coquettish smiles and flourishing them this way and that to disconcerting effect. The testing time comes of an evening, when the Ibiza belle who is thinking seriously of marriage 'interviews' her suitors. Each has his allotted number of minutes, during which time he must plead his case, ardently and eloquently, but with due attention to the clock. Should a suitor exceed his allotted time, the man whose turn it is next has every right to challenge so gross and discourteous a breach of the rules. In the old days the challenge was extended to pistols and knives with a glass of *anis* at some dark hour; and these events have been magnified by the people of Majorca, who still refer to Ibiza as 'that sinister island'.

The Balearics would appear to be the most likely hunting ground for beachcombers in the Mediterranean today, though there are various interpretations and definitions of beachcombing. The Spanish authorities are taking steps to drive out the Hippie from the Balearics, but to the Balearic mind there is a subtle difference between a Hippie or dropout and an unobtrusive beachcomber. You just have to know where to draw the line; and in every Spaniard there is a touch of the anarchist, which is why liberalism rarely seems to work in Spanish territories.

The man who wears a singlet and a patched pair of jeans, grows a beard and has something of the 'scamp philosophy' that has been so charmingly described by Lin Yutang in his *Importance of Living*—well, he *might* just be able to fit himself into the pattern of Balearic life and successfully and even happily live on a mere pittance. At least the harmless type of beachcomber might, not the professional vagrants or political agitators. A beard would be permissible, whereas long hair would be apt to get the beachcomber into trouble, and he would smoke pot at his peril, for the penalty is a long jail sentence. He would be welcomed if he had a guitar, and he might make some kind of a living playing it from *bodega* to *bodega*. To some extent his eating problems would be solved in that, in almost every small café or *bodega*, it is the custom to provide a dish of olives, shellfish and other snacks with each drink purchased. He would have to speak Spanish, and survive better if he spoke it well. You can live on French islands and they will be proud to try and speak English to you; a little French can take you a longer way. Even in the Greek islands a little Greek is often all that is necessary. But in Spanish-speaking countries—no. They do not want to speak another language if they can help it, but speak their own language, use their own idioms, and they can be the most generous people in the world. They will probably shower extra gifts on you. Anyone going to live in the Balearics, especially beachcombers, should master the language first. You can hardly call it Spanish, however, for Castilian or even Andalusian would not be understood here. But it is easy to learn.

Life is not quite as cheap as on the Greek islands, but it is not too expensive if your tastes are simple. Any hankering after luxury, any suggestion that you have 'money to burn' will speedily force up the cost of living for you. The natives here—perhaps it is their Moorish ancestry—are apt to regard anyone other than a simple-lifer as a very wealthy man. It is understandable: they themselves live exceedingly frugal lives and what is quite a small sum to us seems a fortune to them. The result is a tendency to bargain, which forces the would-be resident into the black market unless he really gets to know the language. If you do not, they will pretend not to understand you, to fall back on the phrase *'Mañana por la mañana'*. Tomorrow, as in Spain, is the excuse that comes quickly to the lips, only here *mañana* seems to be the most important and hardest used word in the language.

The Balearics offer much to the really idle escapist, though the authorities may disagree. But, shall we put it this way: if you can honestly escape from the rat race and still appear reasonably law-abiding, always assuming you speak the language, you would be more likely to find peace and happiness in the Balearics than elsewhere. And if you prefer to lunch at 3pm and to dine between 10pm and midnight, then the Balearics are probably your idea of bliss.

Houses and accommodation generally are primitive on the outer islands, though not on Majorca or Ibiza. Sanitation is still not a strong point in some islands. You use oil lamps and cook with charcoal once you get into the remoter parts; charcoal is extremely dirty but excellent for cooking and its fierce heat turns out meals much more quickly than any electric or gas cooker. Water is scarce and each lot of land is given a slope so that rainfall may be drained away and stored for use.

Servants are still easily obtainable and cheap, though not highly intelligent, for hardly anyone seems to have been to school. Their chief virtue, apart from their honesty, is their zest for cleanliness and their care with linen, which they seem to revere. They wash it to perfection and in every house, however poor its inmates, there seems to be an abundance of good linen. The servants are painstaking, efficient and quite often cheap dressmakers into the bargain, even if they are not stylists.

Only a few small islands in the area come on the market, and only occasionally. Consequently their prices are either beyond or just below our set limit. Concentrate on Minorca, Formentera and Cabrera, but do not expect any bargains. For example, Sarganta Island, off Minorca, situated in Fornells Bay, came on the market recently at £90,000 ($216,000) for a mere 6 acres. True, the selling price included 'several old fort-like buildings and its own lighthouse'. Even more astronomical in price was Espalamdor Island, within boating distance of Ibiza, whose 75 acres were priced at £720,000 ($1,728,000) 'as a going concern—all ready to farm, with a successful record of farming over many years'.

If, however, you are content merely to buy a plot of land or a home on one of the Balearic islands, you will find it much cheaper, though, curiously enough, prices for land are higher in some of the more remote islands than on Majorca. On Ibiza you can get a one-bedroom flat for just over £2,000 ($4,800) and building costs here average £30 ($72) a square metre. Plots of land can be bought for £500 ($1,200).

ITALIAN ISLANDS

There are various islands around the Italian coast which from time to time come on the market, but at the moment, in 1972, there do not appear to be many available. A number were sold to foreigners about ten years ago and since then the remainder seem to have been occupied by the Mafia, in some cases exiled by the Italian authorities and in others as owners. An Italian estate agent, who particularly asked for his name to be kept secret, told me in awed tones: 'If we do get an island on our books again in the near future, it will probably be Mafia-owned. But this does not necessarily mean that an enormous price will be asked for it. When the Mafia sell, they like to sell quickly rather than wait a long time for a huge profit'.

Two years ago a number of Mafia suspects were banished to Filicudi, an island off the northern Sicilian coast; to L'Asinara, off Sardinia; and to Linosa, a tiny volcanic island some 120 miles from Sicily and nearer to Tunisia than Italy. Notwithstanding that there are Mafia suspects exiled there, Linosa can be recommended as a remote island well worth consideration. It takes 7hr to reach by sea, and it has only one village, containing some 170 houses painted in red and yellow, cobalt and pink, green and ochre. The village has one shop, one church, a police station, a school and two cafés. There is hardly any tourist trade and a great deal of poverty. Rain falls only twice a year, so there is little agriculture.

To try to buy land in Italy, either on the mainland or on the islands, can be an irritating experience, calling for patience and real gifts of persuasion. Italians who own islands are loth to part with them for sentimental reasons, so bargaining is interminable and it may take years before a deal is finalised. Often, too, the deal is complicated by the fact that Italian islands are owned by whole families rather than one individual, so that while half of them may be prepared to sell, the remainder will delay the deal; and it often happens that one of the owners is an absentee who has gone abroad to live, but who claims that he may want to return to his island one day.

Out-and-out romanticists, to whom a name is everything, may be interested to know that according to a recent advertisement in an Italian newspaper the island of Monte Cristo, 4sq miles in area and made famous by Alexandre Dumas,

Page 123 *Mykonos in the Cyclades*

Page 124 (above) *A wooded corner of Corfu;* (below) *Cephalonia in the Ionian Isles*

can be rented—even for a period of one month. Monte Cristo was once the property of the Italian Royal Family; it is uninhabited except for goats and what is described as 'a serving staff', whose services are included in the rental fee.

The Aeolian Isles include Lipari, Filicudi (see p122), Alicudi, Stromboli, Vulcano, Salina and Panarea, and lie between the south-west tip of Italy and the north coast of Sicily. You reach them by boat from Naples, or alternatively by hydrofoil from Sicily. These are happy unspoiled islands, rewarding to any explorer and all of them surrounded by a sea that shimmers like a peacock's wing.

Lipari is as good a centre as any. It possesses a charmingly romantic town perched on a clifftop, with a magnificent cathedral, some attractive cafés and unexpected delights in its small museum, and its countryside alternates between hills and valleys of vines, olives and figs. Nevertheless, it tends to be expensive and one might do better to explore such islands as Panarea and Salina. Panarea has no roads, no electricity and few amenities, but it has become a favourite haunt for a few escapists who have set up homes here. The centre of social life is the yacht club overlooking the harbour.

Three islands omitted from my survey of Mediterranean islands are Sicily, another Mafia-dominated territory, which in many respects is merely an old-world extension of the mainland of Italy, Sardinia and Elba. The last two have been marred by commercial exploitation.

MALTA

Malta has frequently been cited in newspaper advertisements as a Mediterranean paradise, but politics has now made a nonsense of this claim. It is easy to criticise the Prime Minister, Mr Dom Mintoff, but the recent immigrants themselves are much to blame for the threat to take away their tax benefits. Malta was unfortunately invaded mainly by tax-dodgers and grumblers of prewar British vintage, who could not accept that the British no longer had an empire. They were insular, suburban, bad mixers and lacking any real love for the Mediterranean or its peoples. It may be argued that escapists may pursue an imperial dream if they want to, but I feel this puts them outside the category of genuine island-lovers and they are certainly not going to settle down in the Mediterranean.

However, if Malta is out, there is still Gozo, or Calypso's Isle, 4 miles north-west of it. It is only a 10p trip from Malta, and the voyage takes half an hour. Gozo is the place for genuine lovers of the Mediterranean, with none of Malta's sophistication. You land at Mgarr and climb the road from its harbour up a steep cliff on the three-mile journey to Victoria, the capital. In its architecture it is completely Victorian, having houses with bay windows and carved stone balustrades. Few visitors to Malta get as far as Gozo, but the trip is thoroughly rewarding. The people are poor but hard-working, the men tending the farms and fishing, the women making lace. There are remains of prehistoric temples and some splendid churches, and the tempo of life is much slower than in Malta. Property and land are cheaper also, and some old town houses, with enclosed courtyards, are available for conversion for as little as £1,500 ($3,600).

ADRIATIC ISLANDS

Twenty years ago I said in *Islands for Sale* that 'the islands off the Dalmatian coast and others in the Adriatic seem too near the Iron Curtain to offer any of the blissful existence which they held for many people before World War I'. But now, with the Cold War having slackened off and stability returning to the Balkans, it may well be that the islands in the Adriatic and especially those off Yugoslavia hold bright prospects for the future. There are many islands, mostly quite small, lying close to the Yugoslav coast and offering sun, cheap living and open-hearted generous natives. Many of these islands are within rowing boat or even inflatable-dinghy range of the mainland. With Yugoslavia intent upon maintaining her own brand of 'democratic communism' as against the Soviet version, some of these islands will ultimately be open to foreign settlers.

NORTH AFRICAN ISLANDS

The islands off the Tunisian coast were, before World War II, little known to the outside world, and it was here that the more unusual type of escapist came. I first got to know them during World War II, when almost every island involved a minor operation by the Allies before the Sicilian invasion could be launched.

The most appealing are Galita and Djerba. The latter, which lies close to the Gulf of Sfax and near Gabes on the Tunisian-Libyan border, was the last refuge of the Berber language in Tunisia, but there is no trace of it now. The natives of Djerba speak French and the modern Arabic, yet they claim to hold allegiance to the heretical Abadite sect. There are several versions of the origins of Djerba, which, unhappily, has in recent years become a tourist resort. It was certainly the inspiration for Gustave Flaubert's *Golden Isle* and it is said that it was also the land of the Lotus Eaters, Circe's Isle, and that it contains Calypso's Cave—though the same story is told about Gozo. Djerba, with its palms and fruit trees, is much more fertile than the arid African mainland only rowing-boat distance away; and its inhabitants, who include a considerable number of Jews, are simple kindly folk, quite different from and far more pleasing than the natives of the coastal towns. Living there has always been cheap and is still somewhat cheaper than in Tunisia, which again has a lower cost of living than either Algeria or Morocco. Property in the form of land or houses is not too expensive. A villa on Djerba of, say, six rooms, complete with garden, well and servants' quarters, will cost only £2,500 ($6,000) compared with £6,000 ($14,400) or more in Tunisia.

If I were to choose an island in this part of the world it would be Galita, near Cape Bon and famed for its lobster fishing. This offshore area of North Africa is perhaps the most sophisticated and civilised of all and, despite fierce pride in their new-found independence, the Tunisians are easy to live with. They have absorbed many of the French virtues, and if you can learn to speak a little Arabic the trouble it will take will be repaid a thousandfold.

Arabs can be led but not driven. Give them a fairly free hand, if you have them as servants, and encourage them to use their initiative. You are expected to have servants, or at least one, if you live among them, otherwise you lose prestige and are regarded as rather mean. If you have a servant, he is also probably engaging servants, though the latter will very likely be little more than children and paid meagrely. But for this reason they understand the whole question of master-servant relationship rather better than you do. Do not follow them from room to room, let them work things out for themselves as far as possible and also remember that their sense of time is not as acute as yours—which should not worry an escapist. They can cook to perfection, if you give them their heads, though they will undoubtedly overfeed you, for they seem to believe that all foreigners have gargantuan appetites. Their own cooking is very good, not only special dishes such as the famous Tunisian *cous-cous*—highly spiced vegetables mixed with peppers, raisins, almonds, eggs and chicken, with pyramidal piles of piping hot semolina—but their stews, especially their fish stews, seasoned with saffron.

Arab servants like gregarious masters. If you do not entertain, they tend to become morose and sulky, even if it means they have less work to do. On the other hand, if you entertain, they become gay and excited and seem to share your pleasure. The cynical will say that this is because they anticipate more left-over food that they can smuggle to their homes, but the truth is that Tunisians genuinely like to see visitors around their masters' houses and they love parties even if they are only onlookers.

The Arab scene can quicken the pulse. Its fête days are full of colour and gaiety, the nocturnal music of lute and drum adding a strangely satisfying accompaniment to eventide, and the company of the Tunisians can be lively and stimulating.

7 THE GREEK ISLANDS

'ETERNAL summer gilds them all' was the Byronic tribute to the clusters of islands off the Greek coasts. They combine the romance of all things Greek with the serenity that an appreciation of the ancient world demands: the riches of their heritage assuage the most jaded palate of learning, while their simplicity of life renews one's zest for living.

Here is everything for which the idealist longs—warm dry summers and mild winters, and a timeless existence where the glories of the past are merged with the dreams of the future into a three-dimensional present. Max Beerbohm wove a sophisticated little fairytale about the pleasure-satiated Lord George Hell who suddenly found the charm of virtue and a simple life when he met an innocent country maiden, and how he bought a mask of nobility to hide his debauched countenance. One feels much the same when one returns to these islands after a long period in an ultra-sophisticated sensation-seeking circle of acquaintances.

To turn to practicalities—if it is inexpensive living you want, then come to the Aegean Sea. One could almost call this chapter 'How to have an island home on a slender income'. The only snag is political: if you are a left-winger, you may object to living under the regime of the Colonels; if you are a right-winger, you may spend your time looking round the *taverna* wondering which group of peasants will form the island workers' council if the Communists spring a *coup d'état* on the mainland. Greece, however, has a habit of coming back to a middle-of-the-road regime and among

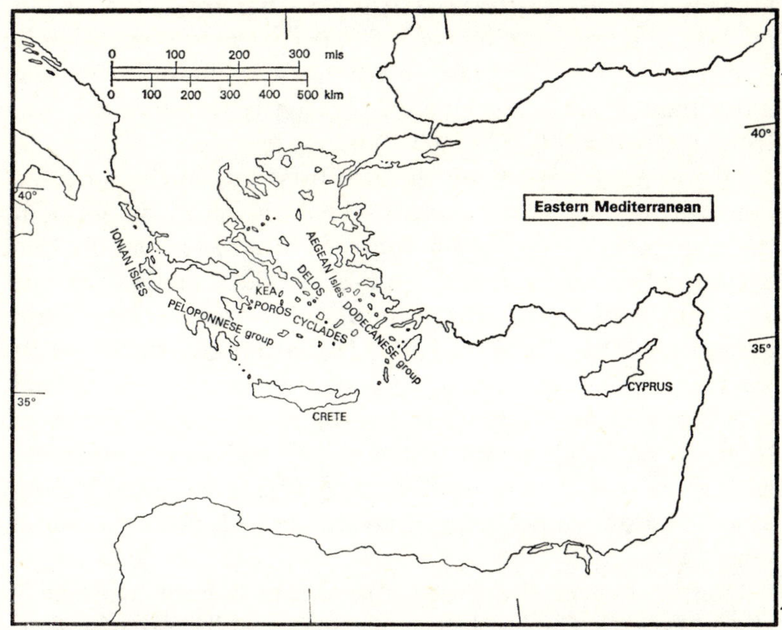

these proud and independent people no tyranny lasts for long.

No fewer than eight islands were for sale early in 1972, ranging from £200 to £25,000. The cheapest was 7 acres in extent and contained some trees and a broken-down but easily repairable small stone quay. That at £25,000 included a habitable villa with all modern amenities, piped water supply from springs, a vineyard of 2 acres and adequate landing-places. For downright romantics there was a bargain in Tourkonissi, a bare rocky island of 10 acres off the coast near Missolonghi, where Byron died. It was advertised for £8,000 and the estate agent's details included: 'Water comes daily by tanker. Uninhabited. Landing-stage'. Also on offer was Lithadonisi Island, 75 acres inhabited by a community of fishermen, priced at £56,250 ($135,000).

There are plenty more in the millionaire's range. One real estate agent, whose clients are said to include Aristotle

Onassis and the Beatles, nearly always has about ten islands for sale at prices ranging from £50,000 to more than £1 million. Indeed Patroclon Island, near Cape Sounion, comprising more than 1,000 acres, though previously uninhabited, was listed at £1,125,000 ($2,700,000) not so long ago.

Islands that appear on agents' lists are much more expensive than those one happens to come across by accident or by making inquiries on the spot. The best way to find a relatively cheap Greek island is to take a holiday in the area and investigate. For example, one island that would almost certainly have fetched £20,000 on the books of an international estate agent was sold privately to a casual holidaymaker for £7,000. Its 15 acres included a stone cottage in good repair, olive and lemon groves and an estate which, with a reasonable amount of work, development and clearing away of bush, would bring in an income of £400 ($1,960) a year.

One or two of the group containing Spetses are nearly always for sale. Curiously enough an island here that was up for sale for £250 ($600) twenty years ago recently came on the market again for £1,800 ($4,320). It has perfectly good well water, a cottage that was partially renovated some fifteen years ago, about 100 olive trees, a splendid beach and some pasture land. The orchard had been allowed to run wild and needed to be restocked. But, my British informant in Athens tells me:

> ... allow some £600 to be spent on repairs and improvements and in three years you could practically live off your island. You would need to keep some chickens, a goat, and perhaps a few pigs, sell the produce of your olive trees and grow some fruit and vegetables. You could catch all and more fish than you wanted.
>
> Olive oil is almost a currency in this part of the world. Your butcher will give you meat in exchange for it and, even if you had no private means, you could live modestly on such an estate. The fact is there is not much you can spend your money on. If you had a private income of a few hundred a year as well, you would have no worries at all.

This estimate is by a sober-minded person who is himself

no simple-lifer. But no escapist likes his escapism threatened by a lack of cash or a bad season. A safe margin for life in these islands, always providing you develop your land, is a capital of £2,500 to £5,000 ($6,000-$12,000) and a private income of £700 ($1,680) a year. If you want to have nothing to do with farming and the running of an estate, even of an olive orchard, then you will need much more.

Greek 'farming' is very different from the arduous business of getting up at crack of dawn on a frosty morning, and keeping pace with modern developments in order to compete with one's neighbours. It is a simple happy-go-lucky business where work in the open air is a pleasure, and where the tending of olive groves and a vegetable garden, watching your chickens and collecting their eggs is very nearly all you have to learn. Make sure that your island, or your estate, has a well in good condition. If you have one, you can be sure that the water is drinkable, since wells are invariably deep.

Your island or estate will provide you with milk, eggs, poultry, pork and bacon, olive oil and vegetables, as well as some fruit. Fish you can catch yourself if you have a boat, but it is not cheap to buy. Meat is apt to be expensive, but not if you have olive oil to give to your butcher. Butter is made on the islands and several make their own wine, though any wine is almost ridiculously cheap. Ouzou, which may be home-brewed, is the chief menace of the islands as far as drinking goes. It is similar to pernod and anisette and is taken both as an aperitif and a nightcap. At sundown it can be a remarkable freshener after a day's work, but taken to excess it becomes a drug, a dangerous habit that leads to forgetfulness and chronic alcoholism.

You can eat and drink well on the islands if you like Greek cooking and the 'roughness' of Greek wine. Like most Mediterranean peoples, the Greeks have a happy knack of making exotic and satisfying dishes out of very little: stuffed aubergines, the mixing of saffron and rice and a few similar culinary tricks may quickly be learned and a study of Greek cookery will pay handsome dividends. You can also be certain of getting a good cook fairly reasonably.

If you want to look for island homes here, concentrate on the group of eight islands in which Spetses lies and on the islands of the Cyclades, some 30 miles from Cape Sounion. Spetses is exceptionally well endowed with a fruitful soil. It has a population of about 3,500. Equally pleasant are the Ionian Isles (see p124), the Cyclades and Aegean groups.

Particularly attractive is Mykonos (see p123) in the Cyclades, a wooded isle with golden beaches that beckon the bather, a row of white windmills on a promontory that makes a perfect setting for the artist, and square houses in Venetian style whose white, pink and blue paintwork gives its town an air of perpetual carnival. You can hire a caique on Mykonos and cross to the Island of Delos, the legendary birthplace of Apollo, where specimens of the architecture of ancient Greece are superbly preserved, thanks to the French School of Athens, which has established a museum here. The remains include the famous Sacred Lake, guarded by its stone lions. The Apollo tradition was so strongly imbued in the island character that it was decreed that no one must be born or die on the sacred island, and those on the point of death and mothers in labour were ferried across to Rhena.

Poros in the Peloponnese has a little port against a background of cypresses and lemon trees where you can enjoy café life and guitars playing tangos until the early hours of the morning. You go to the *taverna*, select a table in the courtyard, order a bottle of wine, and then go into the kitchen to choose your dinner—a pleasant Greek custom. Among an array of dishes and pots and pans you will see piles of vegetables and pyramids of pilaff decorated with red and green peppers, a variety of fish, fried octopuses, chicken and meat. One dines in the silvery moonlight, the night sky like a cloth of purple velvet on which the stars recline like sparkling diamonds, while the three-man band plays its tangos, and the air carries the scent of vine leaves and blue convolvulus that wind round the wooden trellis work. Such a dinner is as ambrosial as any set before the Gods who played and fought on Poros.

Voyaging round these islands need not be too difficult, nor need you wait for all the various connections that normally delay one's travels. The island of Kea in the Cyclades, for example, can be reached direct by powerboat in about 50 minutes from Voulliagmeni Marina, which is only a few minutes' drive from Athens airport. This costs about £2 a head on the basis of five or more people travelling together. Even a prolonged tour of inspection of the islands need not cost much: on Kea you can stay in a room for as little as 27 drachmas a day, which is well under 50p, and in the *tavernas* you can get a varied if simple meal for as little as 17 drachmas. Incidentally, the waiters at the *tavernas* have a custom of writing out any bill that is not understood on the palm of the hand.

The population of the Greek islands has declined in the same way as that of the Hebrides of Scotland: 100 years ago the population of Syphnos was about 20,000 and that of Seriphos 10,000, but now they have 2,000 and 1,000 respectively. Each year more people leave for the cities of the mainland. This decline needs to be arrested if the island way of life is to be maintained, though it does indirectly help the cause of the escapist. On the other hand half the joy of living in the Greek islands is due to the atmosphere created by the islanders themselves, and one would hate to see that dispelled by tourist syndicates replacing the natives.

Fortunately Greek pride will probably prevent any large-scale foreign invasion of tourist promoters. The Greeks not only have a fierce loyalty to their country, but a passionate love for its beauties and heritage. Like the British they dote on islands, and would bitterly resist any large-scale commercial exploitation of them by foreign-owned hotels, casinos and the rest. Athenians, for instance, are apt to keep the name of their favourite islands to themselves: they are frightened of them becoming too popular, especially with foreign exploiters. It is said that every Athenian either dreams of making enough money to buy an island, or guards the secret of the island to which he hopes one day to escape.

In the Dodecanese life is even cheaper than in the other islands, and the price of islands here is almost 30 per cent less than elsewhere. Much the same applies to Crete, which, though ravaged by many wars, seems to retain its ageless quality, probably because its mountains, rising to more than 7,000ft, keep out invading customs if not the invader. The only way to travel into these mountains, and, indeed, through a large part of Crete, is by mule or donkey. In winter the mountains are invariably snow-capped, while in summer, viewed from Canea, their limestone white peaks still give the impression of snow. Here again you find the legends of the Gods that flourish everywhere in the islands, and the peasants will still show you the grotto on Mount Ida where Zeus was born and is said to have been nursed by nymphs.

Canea, the capital of Crete, is picturesque if only for its white minarets, a relic of Moslem occupation. In the sun the city makes a splendid picture, with its sparkling turrets fading into a rose-pink glow at sunset, but in winter it can seem bleaker than anywhere in the Mediterranean. Crete's rainfall, which lasts from November to March, can be extremely heavy.

Though the Greeks will quickly detect the would-be exploiter of the islands and do their best to fob him off, probably by demanding an impossible price, they are sympathetic to the genuine island-lover who just wishes to make a home among them and likes the Greek way of life. Their favourite people are still the British, which is surprising when one considers all the acrimony over Cyprus, and there is no doubt that Byron was the best public relations officer Britain ever had in these parts. When I asked an Athenian friend about the best method of finding an island, he wrote:

> Tell anyone interested to write to the Mayors of some half a

> dozen of the major islands. They should tell the Mayors they are British and they will be absolutely thrilled at the idea of having them as neighbours. They will move heaven and earth to find an island and, if they can't, they'll certainly offer several estates. You've got to remember this—the Greek is intensely proud of his country and he feels prouder than ever if other people want to come and live there. He takes it as a personal compliment. If the applicant isn't British, then he had better compensate for it by flattery.

I pointed out that my friend had first made this statement more than twenty years ago. 'It still holds good,' he said. 'The British are still the best-loved visitors to Greece.' If you think this is an altogether too simple approach, then try one of the estate agents dealing in Greek islands mentioned in Appendix II. Here are some of the very latest deals.

Of six islands in the Ionian Sea three were offered at prices well in excess of £100,000, Sarakiniko, 450 acres, being priced at £300,000. However, two islands were well within our range and both were listed by a London estate agent. One was Sophia (£35,000; $84,000), which lies in a small group of islands and is protected by them—but it had no water on it at the time it was advertised. Vromona, 200 acres and lying six miles south of the main group, cost only £25,000 ($60,000), and had water supplies from a container.

Estate agents' prices for Greek islands often seem to fluctuate so wildly that it is almost impossible to find a norm. For example a 15-acre island off Voula Beach, Athens, was priced at £480,000 ($1,152,000) a short time ago, despite the fact that it had no services and not even a water pipeline, while 200 acres of land on Hydra, which had water, electricity and telephone services, was offered at £70,000 ($168,000). At least two other islands not far from the Greek mainland and not more than 100 acres in size have been listed at £300,000 ($720,000) and £500,000 ($1,200,000) respectively.

So my first advice still stands. Go out there yourself and look round for the real bargains. You will find that the Greeks are among the most likeable of Mediterranean peoples. If they take to you, they will go to immense pains to find the kind of island you want at a reasonable price, even to the extent of taking you out in a boat and showing you several. Nowhere else in the world will you get such co-operation. But you should make it clear what your cash limit is, and, if you are wise, you will make this limit about £5,000 less than you are prepared to go to. For they will inevitably tell you that if only you could spend another few thousands they know an absolute gem....

8 THE INDIAN OCEAN

THE SEYCHELLES

GENERAL 'Chinese' Gordon, that great student of the Bible, persisted in claiming that a fertile valley of gigantic palm trees in the island of Praslin in the Seychelles group was the site of the Garden of Eden. How he arrived at this decision I have no idea, but the legend persists and the few inhabitants of Praslin who have heard of General Gordon repeat the story. It sometimes means an extra tip from an occasional visitor, who is invariably shown the 'forbidden fruit'.

They say in the Seychelles that the 'forbidden fruit' is the only fruit ever to be nationalised. It is the coco de mer, whose trees soar as high as 100ft and the nuts grow right at the top, which suggests that Eve must have been air-minded before Adam thought about it. Some of the trees are 800 years old and are greatly revered by the natives. The Praslin valley was taken over by the Government to save the crops.

You cannot escape the eternal coco de mer and the legend of Eden in the Seychelles. The inhabitants use the 20ft long leaves of the trees for making straw, while string and ropes are made from the epidermis of the stalks. The fruit itself has remarkable properties, as again the islanders will take great pains to tell you. Dr Bradley in his book, *The History of the Seychelles,* says that the coco de mer was washed up on the shores of the Maldive Islands centuries before the Seychelles were heard of, and given the name coco de mer in the belief that it grew from a tree on the ocean bed. In

India a cult was formed to worship the fruit. Doubtless due to its resemblance to the female organ it was credited with wonderful properties. In size, shape and appearance, down to the tufts of fibre-like pubic hairs, it looks exactly like a naked female pelvis, and the delicately flavoured, whitish-green jelly contained in the nut was used as an aphrodisiac. The inhabitants say that its appearance must have dismayed the puritanical Gordon and that doubtless to him it was indeed 'forbidden fruit'.

The Seychelles comprise some ninety-two islands of which Mahé, the main island (see opposite) being rather less than 1,000 miles east of Mombasa. Indeed, it is to East Africa rather than Britain that the Seychelles look for their future prospects. The islands lie outside the cyclone belt and, despite the close proximity of the equator, the climate is agreeable. May to October is the pleasantest time of the year, since from November to April the north-west monsoon blows fitfully, bringing with it heavy rainfall and a stuffy atmosphere.

Most of the population is employed on agriculture and fishing. There are still possibilities for development of agriculture in outlying islands, and land can be leased to immigrants who have the necessary cash. Labour is, however, unstable and plantation labour is apt to move from one area to another. Only in comparatively recent years have islands and properties in the Seychelles come on the market, as previously lack of communications and transport tended to discourage would-be settlers. But with brave attempts to open up the islands to tourism and the creation of an airport in Mahé, the Seychelles are now closer in flying time to Europe than the West Indies. Early in 1972 moves were made to turn the 525 acre North Island, rising 700ft out of the sea, into a holiday development area: it was bought for £150,000 ($360,000) by a syndicate and no doubt the price was enhanced by the fact that cattle and pigs thrive there, as do vegetables and, of course, the inevitable coconut.

The islands range from small flat sandbanks containing a

Page 141
Some of the Seychelles Islands viewed from Mahé

Page 142 *The Veevers-Carters at their home on Astove Island*

couple of palm trees to hunks of high granite mountain, the vast majority of them uninhabited except for exotic birds and tortoises. The Seychelles were for many years none too healthy—hygiene was primitive even for a British colony—but in recent years the situation has been considerably improved, and tuberculosis and leprosy have been largely eliminated. Water supplies are not always free from pollution and immigrants need to take adequate precautions against dysentery and intestinal worms, both prevalent in the native population.

Most development is taking place in Mahé around Port Victoria, the capital, and its airport. The building of the latter was a major engineering job that involved reclaiming land from the sea, and the first jet flew into the island only in 1971. Here the land slopes gently down from the mountains to a wide bay on the north coast. Many plots of land are available in the area of the Casuarina Hills, close to the botanical gardens and the two mountain rivers. Plot prices are about £1,000 a half acre and building costs are about £5 ($12) per sq ft. A small house would probably cost about £6,000 ($14,400) to build. British as well as Seychelles firms are building houses here.

In the early 1970s there were about twenty islands up for sale in the Seychelles, and as there are ninety-two islands altogether, the market should remain good. Silhouette Island, 12 miles from Mahé, comprising 4,000 acres and rising to a height of 2,600ft above sea level, was one of the larger islands up for sale some time ago at an unscheduled price. It has good landing sites and anchorage and is said to contain the only natural forest left in the Seychelles. It has a population of about 700 people, mainly engaged in the production of copra and coffee. Its chief attractions are its extremely rich soil, freedom from sandflies, and excellent big-game fishing. Rather more modest is Cousine Island, 68 acres of palm trees, yielding about 10cwt of copra a month, enhanced by splen-

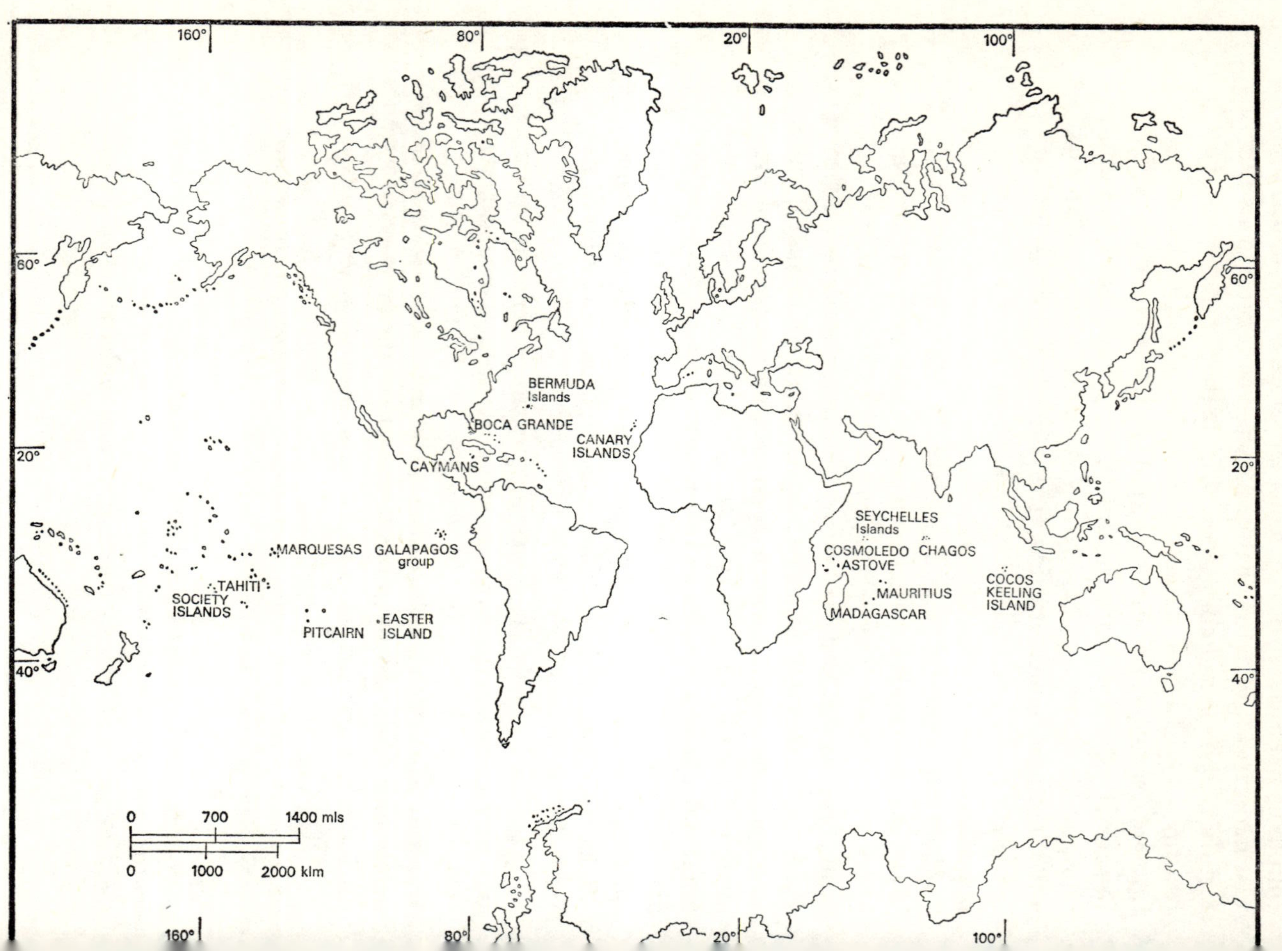

160°
80°
20°
100°
60°
20°
40°
BERMUDA Islands
BOCA GRANDE
CANARY ISLANDS
CAYMANS
MARQUESAS
GALAPAGOS group
TAHITI
SOCIETY ISLANDS
PITCAIRN
EASTER ISLAND
SEYCHELLES Islands
COSMOLEDO
CHAGOS
ASTOVE
MAURITIUS
MADAGASCAR
COCOS KEELING ISLAND
0
700
1400 mls
0
1000
2000 klm

didly expansive sandy beaches, and possessing a natural spring. It was offered on lease at £65 ($156) a month.

If you wish to buy or rent an island, you must be prepared to face intensive bargaining. There has been an attempt by some land exploiters to push up prices on the strength of the building of the airport on Mahé, and some of them are simply trying their luck by asking extortionate sums. You should disregard Bahaman prices in the Seychelles and be prepared to go out there and scout around for one of the smaller islands. The scenery is astonishingly beautiful, more so in many respects than in the Pacific. There are about six women to every one man in the islands, which may explain why one writer on the Seychelles states 'rape is unknown and unnecessary'.

There is a happy touch of Gallic influence in the islands, which gives them an exotic quality and, in some instances, adds a touch of creole beauty to the Seychelloise. The French first annexed the islands in 1742, intending to establish secret spice plantations and break the Dutch monopoly of the spice trade. To keep the secret the French burned the plantations to the ground in 1778 when a ship flying the British flag entered the harbour, but, to the mortification of the crop-destroyers, it was a false alarm: the ship proved to be a French slaver that had hoisted the Union Jack in the belief that the British held the islands. They did not occupy them, in fact, until 1810, when they made them a Crown colony.

Several islands could probably provide one with a modest living from efficient and scientific cultivation. An example of this on a large scale is the island of Fregate, which in January 1963 was put up for sale at £100,000. It is 702 acres in extent, 30 miles east of Mahé, and belonged to Mr Harry Savy, a member of the Seychelles Legislative Council. It had a native labour force of some sixty people and produced copra, citrus and tropical fruits. The owner's residence contained five bedrooms and a living-room, with all modern conveniences; there was fresh running water and the owner was prepared to include in the deal a 15-ton lugger as well as another tiny islet only 2 miles distant.

The prospect excited fifty-five-year-old Mr Leslie Ford, of North Cheam, Surrey, who described himself as 'fed up with white collars, Civil Service routine and twenty years of dull security'. He announced to the *Sunday Express* of 27 January 1963 that he was 'looking for forty-nine people to join him in colonising the island', but, like Mr Stanbury (p59), found that island-buying with the would-be colonisers sharing the costs was somewhat of a pipedream. It is easy to collect candidates, but it is a different matter to find candidates with cash to spare.

Yet Mr Savy's 'kingdom' is probably the best example, though admittedly on too large a scale for most people, of what can be done in the Seychelles. He said:

> Under good management, Fregate Island could bring its owner a net monthly profit of several hundred pounds, exclusive of all its amenities and comfort and the glory of being its uncrowned king. I have cinnamon bushes which produce a few tons of leaves and some bark. We grow just about everything here, including copra and all our vegetables. Oranges, limes, papaws, bananas, cucumbers, pumpkins, koirabi, and even cabbages do marvellously well. We keep pigs, the hens supply us with eggs and, of course, there is plenty of fish.
>
> We crush the sugar cane and the juice, called *bacca*, is highly appreciated by the labourers. It is a nice beverage and is intoxicating when fermented. This drink is to the inhabitants what beer is to the Englishman.
>
> The island produces a monthly average of 60,000 coconuts at present, but this number will gradually increase to over 100,000 in a few years' time. During the past four years some twenty-five to thirty acres have been planted with vanilla and the vines are showing up beyond our expectations. Tobacco grows very well and brings in an appreciable revenue.
>
> Turtles are speared when they come up on the island's beaches to lay their eggs. The waters teem with fish—crayfish, octopus and cockles abound.
>
> And there are legends of buried pirate treasure on the island...

Treasure again! I first heard of Fregate Island, long before it was up for sale, from the late Ian Fleming, creator of James

Bond. He had a passion for buried treasure stories and, having come across some old documents and charts, wanted me to make some inquiries about the Seychelles on his behalf. The documents came from East Africa and they plainly pointed to Fregate Island as the site of the buried treasure, but their instructions were so cryptic (and in parts indecipherable) that it was impossible to interpret them accurately. Fleming went out there to make further inquiries, but the treasure remains hidden. The Seychelles were a haunt of pirates in the eighteenth century and even earlier, for the islands grew fine timber, which pirates used to repair their ships. Some treasure has in fact been washed up on the beaches of some islands, though only in small quantities, and my guess is that the treasure of Fregate Island was probably buried at low tide in a position that is now permanently under water.

COSMOLEDO AND ALDABRA ISLANDS

Also in the Indian Ocean are the scattered islands of Aldabra, Cosmoledo, Assumption, Providence and Farquhar, all belonging to the United Kingdom. Lying midway between the Seychelles and Madagascar, they are certainly unexploited as yet and for a very good reason: few ships travel in this direction and there is no regular air service. To get there you have to make very special arrangements: either you make the long voyage from the East African coast or from Madagascar, or you must—for part of the trip anyhow—charter a plane small enough to land on the islands if, indeed, you can find a safe and suitable landing place.

One enterprising man actually took his family there to settle on an uninhabited island (see p142). Their presence came to light one day late in 1968, when an RAF plane on patrol over the Indian Ocean spotted signs of life on the island of Astove, which was listed as uninhabited. So surprised were the pilot and his crew that they circled low over the island and dropped a facetious message to the occupants, simply saying 'Mafeking is relieved. Happy Christmas'.

This led to the discovery that the occupants of the island were Mark Veevers-Carter, his wife Wendy, and three children with ages ranging from two to nine. They had beached on the island, found palm trees, a roofless wooden house and a graveyard. But no other life. For this was a 'lost' island, nearly 700 miles from Mahé.

The Veevers-Carters put a palm-frond roof on the wooden house and used the verandah as a dining-room. Veevers-Carter was fully qualified to play the role of twentieth-century Crusoe: he had been a Colonial Fisheries officer, treasure-hunter and a schooner skipper before he settled down on Astove. The island had been deserted for years, but he managed to acquire a ninety-nine years' lease of it for the incredibly modest rent of £3.75 ($9) per month! This, indeed, is the kind of bargain one dreams about, but rarely comes across.

There are other similar islands within this area, but one would need to be filled with the same kind of pioneer spirit that Veevers-Carter and his family possessed to survive there happily. They acquired livestock and ducks and cultivated their dream island, ultimately with the help of labourers from the Seychelles. The only fly in the ointment was that Astove was plagued by mosquitoes at night.

'DISAPPEARING' ISLANDS

It is perhaps a 10,000-to-1 chance that you are likely to buy or rent an island that will disappear before you have had time to get there, but this has happened and periodically will go on happening. A seismographical expert assures me that each year a few islands in different parts of the world disappear altogether as a result of earth tremors in the ocean, tidal waves or volcanic disturbances. The danger areas are in the Indian Ocean and the Eastern Pacific. Fortunately, it is usually possible to get reliable data on these areas and so to be forewarned.

The Atlantic Ocean off the Moroccan and West African coasts has had the structure of its bed changed considerably in the past fifty years according to some hydrographers. An Englishman who bought an island off the Moroccan coast some years ago found, on returning there three years later, that only a sandbank remained at low water. It is advisable to choose an island that rises a fair height above the sea, or, if it is low and flat, to make sure that its sandbanks have some adequate protection, such as marram grass, or that it is not liable to constant erosion. Statistics on the tides in the area and a careful examination of changes of current over the years are also well worth studying.

Some tiny islands off Mauritius and in the Seychelles have also disappeared for a few years and then reappeared, to the confusion of navigators. During a cyclone in the Mauritius area 110 years ago, an isle (now called Barly Island) appeared for the first time at the entrance to Port Louis harbour, Mauritius. It is still there and is used as a quarantine station for animals.

In January 1945 one of the small islands of the Carcados group, 250 miles north-east of Mauritius, disappeared for a few days as a result of a terrific cyclonic swell. The islands of this group and others, like the Chagos Archipelago, 1,200 miles north-east of Mauritius, are only a few feet above sea level, which explains their disappearances. The Carcados islands are not inhabited as a rule, but at times groups of fishermen use the main island, St Brandon, as a base for fishing and drying fish. During the 1945 cyclone even the main island and its buildings were swamped by the sea and most of the trees were uprooted. Fortunately a few trees resisted the storm, enabling the fishermen to seek refuge in them during the twenty-four hours of cyclone and heavy rains.

Expedition Island, discovered by Tasman, is now 42ft beneath the ocean, while Torca, an inhabited island in the

Indian Ocean, disappeared in a sheet of fire in 1683. Occasionally a new island is born through earthquake or volcanic action. In 1949 such an island came into being not far from Tahiti. The pilot of a plane flying across the Pacific at the time noticed what appeared to be a puff of smoke on the ocean, and, thinking it might be a ship on fire, flew low to investigate. Then, out of the 'smoke' he saw an island emerge like a giant fish from the sea. As more of it rose above sea level, a volcano could be seen belching sulphurous clouds. The pilot fixed the island's position on his chart and named it after his plane.

CHRISTMAS ISLAND AND THE COCOS-KEELING GROUP

One of the most inaccessible of outposts is Christmas Island in the Indian Ocean: 12 miles long and 9 across, this island with a population of some twenty or so white families is 190 miles south of Java and some 900 miles north-west of Australia. On most maps it appears as just a speck in the ocean.

Christmas Island is really the summit of a submerged mountain 15,000ft high, 1,200ft of which rise above sea level. It has a pleasant climate—always warm but rarely too hot—is luxuriantly clothed with tropical trees and flowers and has none of the pests that usually spoil the traditional tropical island. It has neither malaria, insect plagues, wild animals, snakes, nor rats.

Discovered in 1688 by the explorer Dampier, it was forgotten until 1889, when it was rediscovered by Sir John Murray, the naturalist, and Captain George Clunies-Ross, whose grandfather was 'King' of the Cocos-Keeling Islands, 550 miles away. These two men founded their fortunes at Christmas Island. Sir John was impressed by the rich phosphate deposits and, with Captain Clunies-Ross, formed the Christmas Island Phosphate Company, with a capital of £360,000 ($864,000) in £10 shares and a ninety-years' concession. In October 1948 it was announced that the company had been bought out by the Australian and New Zealand Governments for £2,750,000 ($6,600,000), each £10 ($24) share having increased in value to £76.37 ($183).

The island's population today is about 1,000, entirely imported. Phosphates constitute its sole industry and it is estimated that it probably holds more than 3 million tons of this valuable commodity. Flying Fish Cove, at the northern extremity of the island, is its only landing-place. Immediately before World War II the island enjoyed a boom; it had a population of about 1,500, with such civilised amenities as a cinema, a hospital, company shop, a mosque for the Moslem inhabitants and a temple for the Chinese. Just before the Japanese invaders landed here in 1942 a 'scorched earth' policy was adopted: excavation tools and plant were smashed, and the Japanese were unable to continue phosphate production. But within a few years of the end of World War II the island was restored to normality.

In October 1949 the Christmas Island Agreement Bill was passed in the Australian Senate. The island itself was then an administrative dependency of Singapore, but under this agreement the Australian and New Zealand governments jointly bought out the company and established a commission to run the concern. Then in 1958 Christmas Island became solely an Australian possession. Anyone who thinks of settling there and has suitable qualifications should apply to the Phosphate Commissioners, New Zealand House, London, who are the managing agents.

This island should not be confused with the other Christmas Island between Fiji and Honolulu, which was bought by the British Government from a private company for £50,000 in 1950—as a nuclear test area! Escapists should remember that a number of remote and uninhabited islands are in danger of being exploited in this manner by one or other of the big powers, notwithstanding the warnings of ecological experts. The US Government, for instance, carried out an underground nuclear test at Amchitka Island in the Aleutians in November 1971, despite warnings of the hazard of starting a chain of earthquakes.

Few people could say straight off where the Cocos-Keeling group of islands lies. They should not be confused with Cocos Island in the Pacific. That island is supposed to contain treasure. The story goes that certain Peruvians, panicking when revolution threatened 250 years ago, packed all their gold into the schooner *Marie Dyer*, whose crew mutinied and sailed off with it, only to be wrecked on Cocos Island. Sixty-four expeditions have failed to find the treasure, reputedly buried there, one of the expeditions involving a diplomatic dispute between the United Kingdom and Costa Rica in 1934.

The Cocos group—there are twenty-seven of them, some no more than atolls—is situated in the Indian Ocean, 1,280 miles south-west of Singapore. They were discovered in 1609, but the first European to settle there was Alexander Hare, a swashbuckling adventurer who came from Cape Town in 1826. With notions of establishing in the islands a tropical Utopia, Hare took with him a large harem, a score of male slaves (presumably eunuchs), several labourers and a band of musicians. The number of women in Hare's harem has been put as high as 1,000, but it was probably less than 100. Statistics show that in 1829 Hare's entourage numbered only ninety-eight—thirty-six men, twenty-five women and thirty-seven children; and in 1836 the population of the islands was only 155, including newcomers.

Hare, who had already scandalised settlers in Borneo and Cape Town by his uninhibited pursuit of women, believed that only in the Cocos Islands would he find lasting peace away from hostile and envious neighbours. But such tranquillity as he did find was rudely shattered the following year when John Clunies-Ross, a Scotsman, and his wife arrived. Hare and Clunies-Ross agreed to share the islands, but the arrangements proved unworkable. Clunies-Ross, a dour and puritanical man imbued with Episcopalian ideas of propriety, disapproved of Hare's way of life.

A crew of Scottish seamen who arrived with Clunies-Ross also disapproved of Hare, but for a different reason. They believed his women should be shared among them and they took to raiding his harem in the hours of darkness. Hare tried to buy them off with bribes of roast pork and rum, but when this failed, withdrew with his remaining women into a stockade he built at Pelau Bras. To this day this place is known as Prisoners' Island.

The seamen's raids, however, continued until, in 1831, the last of Hare's women had gone. Yet, still indefatigable in his quest for amorous conquests, Hare left for Batavia, where he started a new harem of oriental beauties. In 1857 the Cocos-Keeling Islands became part of the British Empire and in 1886 Queen Victoria formally ceded them to the Clunies-Ross family, which initiated a programme of progress that brought almost uninterrupted prosperity and peace to the islands for the best part of a century.

Life there is still as peaceful as you will find it anywhere. Crime is practically unknown. There are few laws, but the most important is that anyone who commits a crime shall be banished to civilisation! That threat is enough to keep the inhabitants law-abiding. Immigrants are not in the ordinary way accepted, but there is from time to time scope for the services of someone from the outside world, and there are always a few of the smaller islands that might be available on lease.

The Cocos-Keeling Islands have all the advantages of a warm climate without the soul-destroying heat of the tropics. Sea breezes keep them cool and the seasons are generally very much alike, except during the monsoons, when disastrous cyclones have occasionally caused great damage. Only three of the islands are at present inhabited. Most of the people have descended from Javanese, Chinese, Malays and South African kaffirs. There is work for all males over the age of fourteen and for as many females as wish to take employment. Workers can retire on half pay at the age of sixty-five, but the relatively high standard of living for an island race has produced a sturdy healthy people, and most islanders prefer to continue working after that age. They are free of most infectious diseases, including the common cold, and anyone who leaves is never permitted to return, a factor which is of some importance in maintaining the island's health record. Thus none of them have any knowledge of the outside world—again, no doubt, an aid to contentment.

Industrial disputes are unknown. A three-roomed house and furniture is given to each newly married couple. There is only one rule for marriage: the man must be at least eighteen and the girl sixteen. Free medical attention and communal care for widows and orphans are provided. The Scottish influence of the Clunies-Ross family is apparent: not only do many of the islanders bear Scottish names, but one of the village thoroughfares is called Sauchiehall Street. The home of the Clunies-Ross family, Oceania House, was built in 1836 of stone imported from Scotland.

These twenty-seven coral islands vary in length from several miles to a few hundred yards, lying in a horseshoe formation around a limpid ultramarine lagoon. Nowhere do they rise more than 20ft above sea level, yet there are no reports of any of them having ever come within the category of 'disappearing' islands. The principal island is Home Island and the other large ones are Direction, Horsburgh, South and West.

In 1955 administrative control of the Cocos-Keeling Islands passed from the United Kingdom to Australia, and an Australian air base was established. The only conditions attached to the transfer were that the native population should have the opportunity of becoming Australian citizens, that the Clunies-Ross family should retain ownership of their property and that British aircraft should be allowed to use the air base.

One man has discovered the best of both worlds by commuting between the United Kingdom and the Cocos-Keeling Islands. Mr James Dixon has a bungalow home at Leadenham in Lincolnshire, where he spends four or five months every other year, while the rest of his time he lives and works in the islands. He is manager of the Clunies-Ross Estate. Originally he went out there as technical watchkeeper for the Cable & Wireless Company, but after a year resigned to take up his present job. Like the 400 islanders the Estate Company employs, Mr Dixon enjoys free housing, electricity and medical attention, pays no income tax, and all the goods he buys are duty free. Prices are kept low by Estate subsidies. While copra is the chief product of the local economy, the Estate Company has a stake in shipping interests, merchandise of all kinds and boat-building. Mr Dixon's job, in fact, often leads him to use a sailing boat—a jugong—to go about his work.

In September 1972 an era ended: Mr John Clunies-Ross announced his abdication and conceded Australian sovereignty over his islands. Provision has been made for an elected chief executive to administer local affairs. Both Christmas Island and the Cocos-Keeling Islands remain tax havens.

9 NORTH AMERICAN ISLANDS

CURIOUSLY, the area in the world with the most islands is one which is rarely mentioned—the North American continent. No one has investigated these islands more thoroughly than Ralph Froman, the American poet and author, who travelled thousands of miles cataloguing islands for sale or to rent in sea, lake or river from Alaska to Florida. Mr Froman carried out his research in the early 1950s, but the major part of his research is still valid. The most important and fascinating discovery he made was that there were more than 1 million islands and islets in North American waters and that up to 100,000 Americans owned or lived on islands. He wrote:

> I discovered, conquered and settled briefly on my first island at the intrepid age of eleven. It was about thirty feet long, half that wide, and surrounded by the two-foot-deep, summer-stagnant waters of the Boise River near my home town of Caldwell, Idaho. After dislodging the aboriginal muskrats, I had a fine Robinson Crusoe time until my parents found me out and decided that even two feet of water was enough to drown in.

He also discovered that after World War II islands could be bought for as little as $17 or as much as $25,000. The results of his travels appeared in his book *One Million Islands for Sale*, which, unfortunately, is now out of print. That the demand for islands is insatiable can be adduced from the fact that some years ago when the United States Government advertised six islands for sale there were 2,500 eager inquiries. Curiously enough, the predominant profession among the applicants was that of dentist. Mr Froman's price range of $17 to $23,000 must be multiplied by ten to reach today's

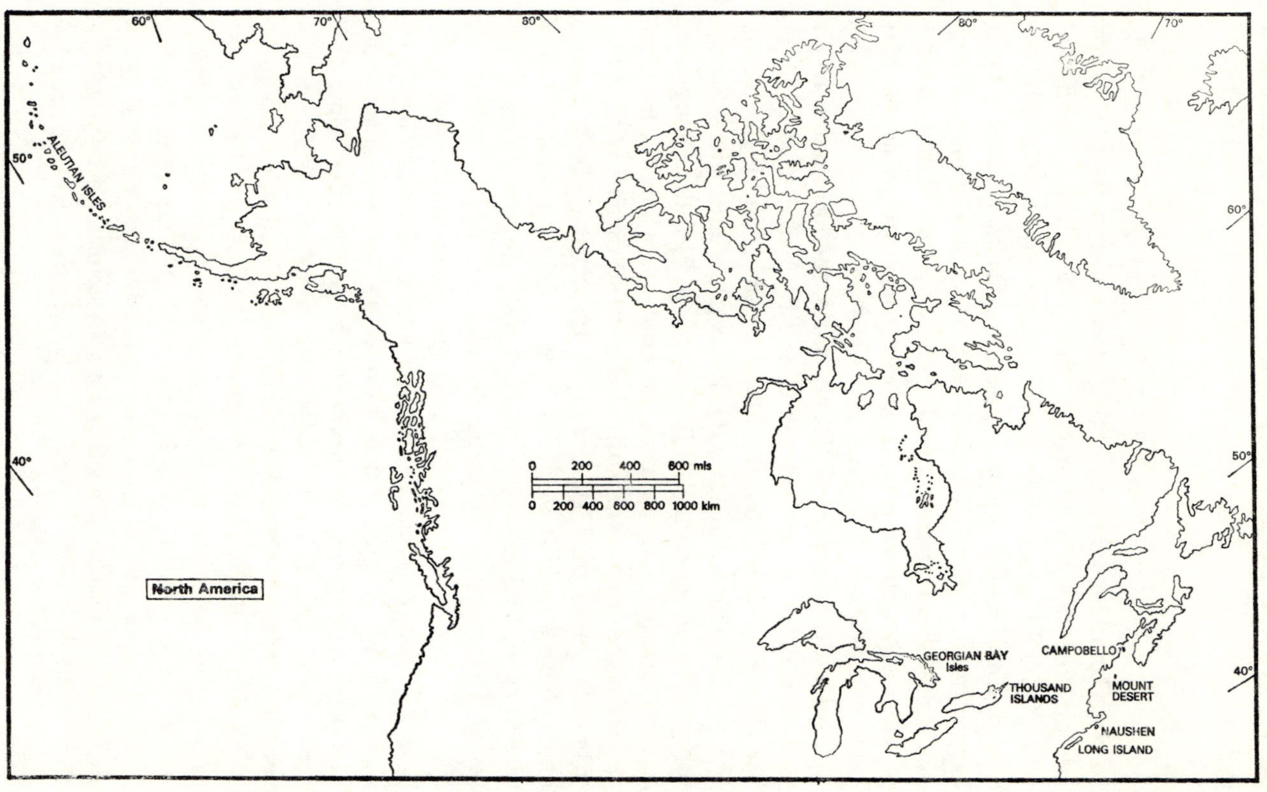
North America
ALEUTIAN ISLES
GEORGIAN BAY Isles
THOUSAND ISLANDS
CAMPOBELLO
MOUNT DESERT
NAUSHEN
LONG ISLAND
0 200 400 600 mls
0 200 400 600 800 1000 km
60°
70°
80°
80°
70°
60°
50°
40°
50°
40°

prices, but North America still has the widest range in island prices.

CANADA

In the Canadian lakes and rivers there are many islands for sale or lease that would attract the farmer, hunter or fisherman. Ontario, more than any other province, is richly endowed with islands (see p159) and an official of the Ontario Government reports that there are 'some hundreds of thousands of islands here, mainly in the many lakes of the province, though some of them are only accessible by seaplane or canoe'.

A splendid example is L'Ile-aux-Oies, the local name given to Bonaventure Island off the Gaspe Bay coast, east of Quebec. The French name is used because the island is the haunt of what are known in the vicinity as gooselight birds (something akin to gannets). The island is 2½ miles long and attracts boatloads of tourists in summer, when its hotel is open. At one time ownership of the island carried with it the title of Seigneur de L'Ile-aux-Oies. The strongly built manor house there was designed to keep out the bitter winter winds as well as the Indians, but the thick walls did not prevent the savage Iroquois from scalping the Seigneur and his wife some 300 years ago. One Seigneur sold the island to the nuns who founded North America's first hospital in Quebec City, and was paid in playing cards signed by the Governor and stamped with the Royal Seal, which were at that time the currency of New France. In 1960 the nuns offered this make-believe feudal estate and profitable farmland not for a pack of playing cards but for the sum of £100,000 ($240,000).

For those who like a northerly climate it is worth noting that there are about 1,700 islands in the St Lawrence River. They all have possibilities for the person who likes an open air life with plenty of fishing and shooting and appreciates their wild but entrancing scenic beauties. Known as the

Page 159 (above) *Typical summer cottage on an island in Bashkung Lake, near Minden, Ontario;* (below) *Some of the Thousand Islands group in the St Lawrence River*

Page 160
Miss Sniechowski on her home-made island in a Suffolk river

Thousand Islands group (see p159), they extend from the eastern end of Lake Ontario, not far from the Adirondack Mountains, far down the St Lawrence River. Most of them are in Canadian territory, but about 600 are in United States waters. Normally there are some few score of these islands for sale or for rent on the books of the estate agents in New York and Ottawa.

One of the Thousand Isles is known as Heart Island and in its name lies a bizarre story. George Boldt, who owned the Waldorf Astoria Hotel in New York, bought the island as a gift for his wife and actually altered its shape until it resembled a heart, filling it with ponds containing swans. When she died, he was so grief-stricken that he left it never to return, his one monument being a castle he had built there as a miniature replica of the Waldorf Astoria.

Still further north are the many islands off the coasts of New Brunswick and Nova Scotia, totalling about 1,800. The most popular of these is the New Brunswick group in Passamquoddy Bay, one of which, Campobello Island, was once the summer retreat of President Franklin D. Roosevelt. The Nova Scotia coast is somewhat bleaker, but there are many islands to choose from along 300 miles of Atlantic seaboard. The most sheltered of these are among the groups in Mahone and Tuskey Bays.

On the whole Canada, possibly because of its much smaller population, but also because of its national and provincial government policies, offers much better value than the USA. The Canadians have preserved islands in their natural state and kept prices within reason, taking only fair profit margins. Until the end of 1970 the Ontario Government, which actually controls a majority of the islands, rented and sold them at prices far below those obtaining in the United States. They based prices fairly according to acreage. The renting rate was at one time as low as $20 an acre and the selling price $100 dollars an acre, the only stipulation being that the buyer must spend specified amounts on improvements and developments within eighteen months of the purchase.

However, the Ontario Government has now stopped these deals on the ground that the Lands and Forest Department was not happy about the development of Crown land for private use. Their policy now is to retain the islands entirely in their own hands and to allow as many of them as possible to be given over to the public for picnic and camp sites. This, of course, does not apply to the many islands already in private hands.

According to a careful check made with estate agents in Ontario in 1971 a 2 acre island in the Thousand Islands group, without any buildings on it, recently changed hands for $8,500 Canadian, while a 3½ acre island in the same area, with a large stone and frame house, went for $20,000. The Georgian Bay Islands come up for sale with rather greater frequency. Two examples at the time of writing are a 7 acre island with seven cottages some 15 miles from Perry Sound (140 miles from Toronto), up for sale at $42,000, while a 2½ acre island in Moon River, with only a fishing shack on it, is expected to fetch $19,000. A slightly larger island nearby, with a cottage, a sandy beach and a stretch of woodland, was sold for $39,000. The smaller islands tend to come on the market rather more often than the larger ones, with the result that prices for the latter tend to be very high. A 220 acre island complete with a shooting lodge and three cottages recently went for $200,000.

It is only in the silence of the northern Ontario lakes that a few genuine bargains remain. A 46 acre island lying across the water from the old goldmining town of Red Lake, Kenora, was advertised for sale at $6,840. There are no buildings on this island, but a 3 acre isle in the same area, possessing a fishing lodge and five cottages, was sold in 1971 for a mere $14,000. However, you would need to be a real escapist to seek islands in these outposts, as the Kenora district is 1,100 miles from Toronto.

The *Ontario Journal & Tax Sales Register* has a list of

hunting, fishing and vacation plots, including some islands, that have become available for auction through non-payment of taxes.

ALASKA

If you are tempted by the far north, there are thousands of islands between Puget Sound along the coast of British Columbia right up to Alaska. Should you have a taste for remoteness, then in Alaska you will have to deal with the United States Forest Service and secure their approval before you can make a deal. If you are lucky, it is possible here to rent an island of 500 acres for a mere £50 ($120) a year. But Alaska can hardly be described as a paradise: the climate alternates between a steady downpour and fog, and many islands are mere rocks, though some are kept relatively warm by the Japanese Current. Most are used as fur farms.

NEW YORK

Mr Forman said in the *New York Times* of 14 June 1953:

> One has to be rather lucky and willing to spend a good deal of money to find an island within commuting range of New York City. There are several hundred islands scattered among the lagoons of Long Island's south shore, along the north shores of Long Island Sound and in the lakes of Connecticut, southern New York State and northern New Jersey. But the more desirable ones are quite expensive and, when rentable, usually spoken for well in advance.
>
> Between the mainland and the long, low semi-urban islands off the coast of New Jersey are a number of bays and harbours dotted with hundreds of islets. Most are low, marshy spits of sand, but there are a good many comparatively high and dry ones with summer homes. Almost the only way to find one for rent is by on-the-spot exploration. Real estate dealers in the area seem to take little interest in them.

Mr Forman's information is still relevant. I inquired into the possibilities offered by these commuting-range islands and actually traced a New Yorker, Mr Winthrop Shorter, who commutes from his own island daily in summer, spring and autumn. Needless to say he is a fishing enthusiast, but his story illustrates what a little enterprise can do:

> I had a run-of-the-mill, boring desk job in New York when I came back from the Korean War and I found my mind drifting towards an island of my own during those years. The trouble was that I needed an island somewhere within fairly easy range of New York City and none within my price range were ever advertised. Finally I decided to take a holiday and spy out the prospects for myself. Well, the New Jersey coast seemed to be as good a bet as any, as an inspection of maps showed that in this area there were so many islands that it was impossible to believe somebody wouldn't be prepared to sell or rent one.
>
> I had to go rather further than I expected to find something small enough for my pocket. It was, I recall, a very wet day and a local fisherman agreed to take me for a cruise around some of the groups. I was soaked to the skin, but I couldn't give a damn, I was so thrilled. Here were literally scores of islets and the joy of it was that the very smallest lay between the larger ones and the coast, so they had some protection.
>
> I was shown no fewer than twenty-three which, I was informed, could be bought or rented. I was, of course, anxious for something I could afford without burning up all my spare cash. For that reason I didn't want anyone to think I was over-ambitious. So I rather cunningly said, 'Well, all I want is a little sand bank where I can put up a hut and fish.' It worked. I was offered an islet that was rather more than a sand bank: it was about eighty yards long and fifty yards wide, mainly sand, but with some grass, scrub and sea flowers. I saw it at its worst on a day when the rain was driving across it in a searing wind, but I noticed that at one corner of the island there was a steep bank of sand—about twenty feet high—over which the waves did not appear to dash. This high part of the island was on the leeward side. Somewhere below that, in the middle of your island, I thought, you might be able to grow something.
>
> I bought that islet for a second-hand car plus two hundred dollars. The man who owned it had used it for fishing and, as he was crippled with arthritis, he no longer had any use for it. I doubt if everyone would be as lucky as I was. I only bought it in 1965, but you should see the changes now. At weekends I set about building a three-roomed cottage on the island, with the help of friends. We finished it in about three years, including a tank on the roof to trap the rainwater. I have planted some fir trees there and they are coming along slowly. Fir trees are devils to grow; I lost the first three I planted, but have suc-

ceeded with all the others. I grow a few vegetables—nothing spectacular, of course, but this is really a fisherman's paradise.

One is frequently asked whether it is possible to experiment first at relatively low cost to see whether one can adapt to island homemaking. In the North American continent the answer lies in Lake Georgia in New York State. More than 100 islands here are actually owned by the State and these possess about 500 camp sites. If you want to try it out first, you can experiment here. I should stress that you will not have an island all to yourself, though the experiment may well develop the urge to go it alone in future, but for around $70 you can rent camping equipment, including a canoe and tent, and permits for the use of camp sites are free. There is another advantage in trying this out: Lake Georgia has some forty privately owned islands, and it is always possible to rent or buy one of them if the experiment proves to be a success.

FLORIDA AND THE GULF OF MEXICO

Some of the islands off the coast of Florida and in the Gulf of Mexico are much cheaper than in the Bahamas, and this is a happy hunting-ground for the escapist who possesses a modest capital sum. A friend was offered 28 acres of Pine Island in the Gulf of Mexico for £850 ($2,040). That included a plantation of orange, lime, grapefruit and mango trees, and a small building, 12ft by 20ft. He wrote:

> I have a strong feeling I shall finish my life somewhere down here, watching the oranges ripen. The weather is wonderful for nine months of the year and there are few places about which one can say that.
>
> But this is a splendid territory for discovering islands. For two years now my wife and I have spent our holidays on an island in the Gulf, Boca Grande, the former haunt of a famous pirate named Gasparilla. Heavenly spot! The area grows all the fruits above-mentioned and lemons, peaches, strawberries and all the northern vegetables as well. The swimming is incomparable, the

sailing glorious, there is an eighteen-hole golf course and it is quite unspoiled.

In the south and south-east of Northern America the coasts of Florida, South Carolina and Georgia are the principal island areas. Mr Seymour J. Bennett of Miami, who is himself an inveterate 'collector' of islands and has compiled two dozen albums of photographs and information about Floridan islands, claims that there are 'more than half a million islands in Florida alone'. Well, I am prepared to accept Mr Forman's total of a million islands in or around North America, but half a million for Florida alone seems a phenomenal number. An examination of maps and charts does not confirm this, but perhaps if one included in this total every mangrove swamp islet of 30-40sq yd, such a claim might be substantiated. However, the map of Florida shows clearly enough the group known as the Ten Thousand Islands, not very far from Florida City on the south-west tip of the state. Most of these islands are too low and swampy to be of much use, but those near the northern edge of the group are habitable, and some have villages on them. There is a wealth of vegetation from coconut palms to avocado and papaya trees. Windmills are sometimes used to generate electricity. For centuries the Ten Thousand Islands were avoided by sailors, as the whole area was marked as navigationally dangerous on charts. Even today the best method of exploring many of the islands is by poling a punt through the narrow waters, overhung by dense jungle.

At a conservative estimate there are at least 15,000 habitable and worthwhile islands round the coast of Florida, ranging from the Ten Thousand 'mangrove' islands to the coral-reef formations in the Keys.

NEW ENGLAND

The New England and Maine coasts are plentifully supplied with islands of all sizes, probably the best selection being off Maine, where there are 2,500. Best known are the large islands of Mount Desert and Arcadia National Park, off Belfast, Maine, but they spread thickly as far as Portland, notably in Casco, Penobscot and Muscongus Bays. Agents dealing in these islands are to be found in Portland, Camden, Thomaston and Waldoboro.

In the nineteenth century this was the playground and recreation centre of America's wealthy families, and star-crossed lovers, hermits and painters also flocked to the area in quest of solitude and peace. Not surprisingly a whole literature sprang up around these various groups of islands, some based on fact, much of it inspired fiction. Oliver Wendell Holmes, a frequent visitor, wrote thus of Naushen Island:

> Where have I been for the last three or four days? Down at the Island, deer-shooting... The Island is where? No matter. It is the most splendid domain that any man looks upon in these latitudes. Blue seas around it, and running up into its heart, so that the little boat slumbers like a baby in lap, while the tall ships are stripping naked to fight the hurricane outside... trees, in stretches of miles; beeches, oaks, most numerous... Fresh-water lakes; one of them, Mary's Lake, crystal-clear, full of flashing pickerel lying under the lily-pads like tigers in the jungle. Such hospitality as that island has seen there has not been the like of in these our New England sovereignties. There is nothing in the shape of kindness and courtesy than can make life beautiful, which has not found its home in that ocean-principality.

Naushen Island was bought by John Murray Forbes, the railway builder, of Boston, and it has remained in the family for three generations. It is not far from Martha's Vineyard.

AMERICAN LAKES

It is, however, in the lakes even more than around the coasts that one is likely to find a sheltered, reasonably accessible and worthwhile island. Michigan and Minnesota have lakes with many islands. In Georgian Bay, at the northern end of Lake Huron, are 30,000 islands, about 5,000 of which are inhabited. Here you will find homes to suit all pockets, from the fisherman's shack to the large mansion. In New Hampshire, which has many lakes, there is the bleak but magnificent Shoals Group, and inland at Lake Winnipeasaukee in some lush and verdant country are 274 islands, all said to be habitable. Ralph Froman writes:

> ...if you visit the lake on an island quest, some enthusiast is sure to tell you that it has 365 islands, one for each day of the year. This odd fiction is a favourite with the island area boosters all over the continent. Someone recites the claim for nearly every lake or bay with more than a hundred or so islands. I have yet to find one who could back up the claim.

The islands in the lake districts of the North American continent are too numerous to catalogue. The best of them fall swiftly into the hands of the big estate agents, who will supply you with long lists, but such islands are expensive. Lake Champlain, and the Rideau and Nipissing Lakes offer a wide choice. The selling and renting of developed islands, with houses, has become big business. For many years the tendency was for such islands to be let rather than sold, but inflation has encouraged many island-owners to sell at a profit, and, of course, the real estate agents are much more interested in a sale than a let. Those in the far north and north-east are the cheapest, as, perhaps, is only to be expected. The price for uninhabited but privately owned islands in the north-east can be as low as $600, but those already developed, with fresh water supplies and house or cottage, as well as adequate landing facilities, cost between $6,000 and $600,000.

10 MISCELLANEOUS ISLANDS

It may be that by the time you have read this far you are still undecided about your choice for an island home. We have by now exhausted the major groups of islands, so it merely remains to scan the maps and charts for some outposts we might have overlooked.

SWEDISH ISLANDS

In Sweden the possession of an island is almost a status symbol. It is the dream of most inhabitants of Stockholm to earn enough money to buy a 'weekend island' to which they can go in summertime. Fortunately the dream is rather more easily realised in Sweden than elsewhere, because the population is small for so large a country and thousands of islands are scattered round the coast and in the many lakes. Each summer weekend sees a steady trek of cars, laden with camping gear and provisions, heading for various island hideouts. Some Swedes are content just to pitch a tent and spend the time canoeing and fishing; others, more ambitious, build a handsome wooden bungalow on their island, fitted up with all those modern labour-saving devices in which Sweden is so far ahead of the rest of the world.

There are several lakes within easy reach of Stockholm and hundreds of islets round the coasts north and south of it. Midway between Sweden and Finland, at the entrance to the Gulf of Bothnia, is a large concentration of islands. One of them, only 7 miles from the coast of Finland and only 4 acres in area, is appropriately named Timso, which means

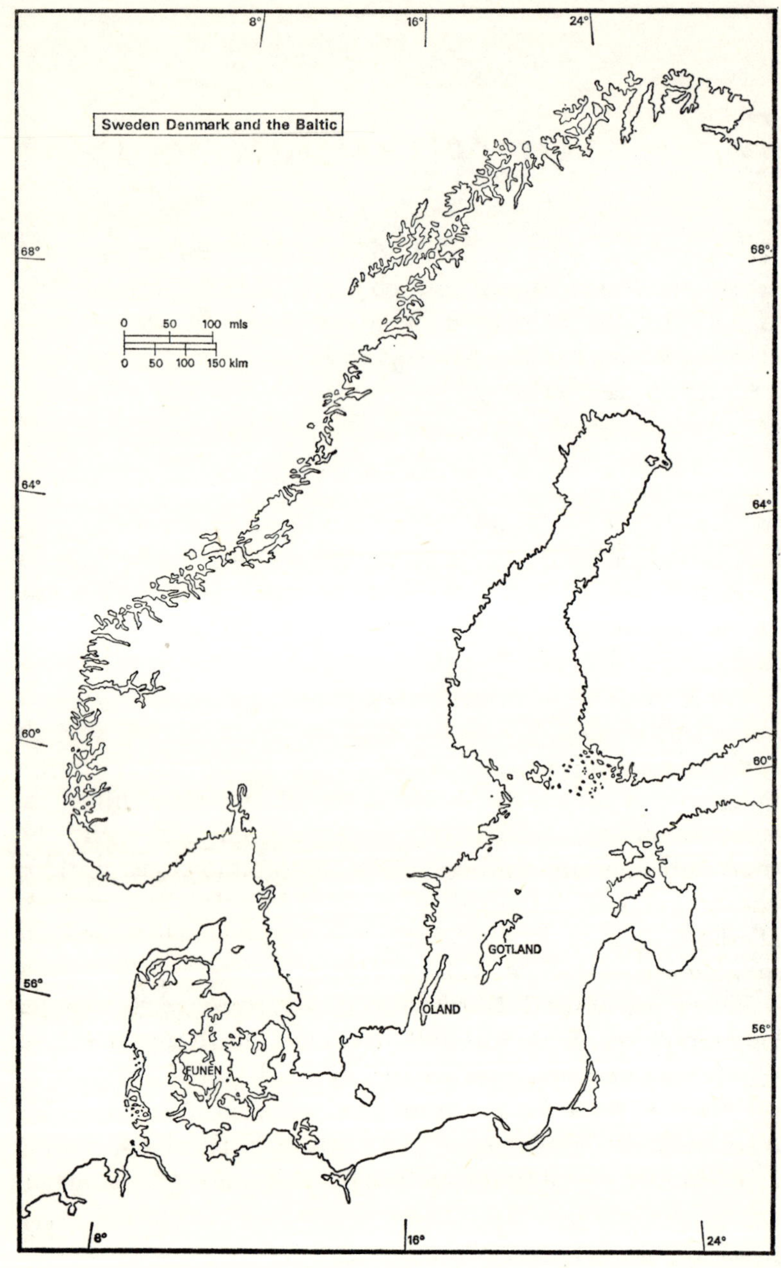

Sweden Denmark and the Baltic
0 50 100 mls
0 50 100 150 klm
8°
16°
24°
68°
64°
60°
56°
GOTLAND
OLAND
FUNEN

'Tim's Island'. I say appropriately because it is owned by Tim Andrews, an American professor who lectures at the university in Helsinki from Monday until Wednesday and then spends the remainder of the week on his island. In summer this is pleasant enough: it merely involves a 60 mile journey by road and a 7 mile sea trip. But in winter it sometimes means skiing across the frozen sea. Still, hardy Tim Andrews has been doing this for about twenty years. He says:

> I'm just crazy about islands. I started off by renting a tiny part of one and lived there for two years in a sauna bath hut. Then one day I was exploring around in my boat when I ran into a rock. I couldn't get the damn thing off, so I thought 'what the hell, I'm in the water anyway, I might as well swim over and have a look.'
>
> And that's how I found this island. Oh, God, it was so, so tremendous, I just couldn't tell you. I just had to have it. I found out who owned it and then began the wooing . . . it was like trying to get an heirloom from an old spinster. They just wouldn't let go.

But Tim Andrews was persistent, and in the end they agreed to sell the island to him for £800 ($1,920). He had to wait nine months for Government approval and then he built himself a house. In Scandinavia building yourself a house is rather easier than it sounds. He drew up his own plans and sent them to a firm specialising in prefabricated building. They worked it all out, cut the logs to the right shape and size, all numbered, and sent them to him with a construction plan. All he had to do was to put them together. He runs his cooker and refrigerator on Butane gas and burns oil lamps. 'All summer long I can just fish and read and sail.' And in the winter? 'I've got a huge log fire in the lounge and those marvellous Finnish stoves in the bedrooms. I've got all my stores and plenty of food. My mail is delivered, there's a cow on an island close by for milk, and I melt chunks of ice for my water. And I ski off and on the island. Absolutely no problems.'

Well, not every island enthusiast might be so capable of

revelling in the hard life as Tim Andrews. But for those who favour a northern clime the islands in the south of Sweden, within easy reach of Stockholm, offer splendid possibilities, and many of them are equipped with most of the requisites of civilised life.

Visit Oland, off the south-east coast of Sweden. It is some 90 miles long and 10 miles across at its widest point, and a paradise for the archaeologist—there are many sites from the Stone Age onwards in its wide open spaces. There are also 430 windmills, none of them working now, but their buildings are used as cafés or converted private houses—and they make admirable dwellings. The whole area is carpeted with wildflowers in summer. Oland is only 25 minutes' sailing time from Kalmar on the mainland.

Another attractive large island is Gotland, which possesses in Visby probably the most attractive and picturesque town in the whole of Sweden. It is surrounded by forty-four towers and is filled with the kind of high-gabled old houses one sees in books of fairy stories.

Much information on Sweden's islands can be obtained from the Swedish National Travel Association, which has branches in most capitals of the world.

DANISH ISLANDS

Further south in Denmark there are much warmer islands. South-west of Zeeland is the Funen Archipelago, a string of low wooded isles where the blue Baltic laps lazily against golden sands and few sounds break the silence but the cry of the gulls and the chugging of passing fishing-cutters. On the island of Lyoe, 2 miles long and 1 wide, one can rent a furnished cottage and combine the balm of comparative solitude with congenial companionship. There are only 150 inhabitants of Lyoe, two shops and seven duckponds, but the scenery is delightfully wooded and the people are gay and charming, with the ever-present sense of humour that so typifies the Danes.

There are many other islands in the vicinity—some with homes and some without, some for rent and some for sale. In any event these islands offer plenty of scope to those who want island holidays. To rent a typical four-bedroom cottage, with all amenities provided, including crockery and utensils but no bed linen, and with all services laid on, costs £16-£25 a week.

If you want to explore the islands, you can indulge in a tour known as 'Funen on Two Wheels', which consists of a cycling tour of the islands, sleeping and eating at good hotels. At the yachting centre of Svendborj you are given a map, an itinerary, vouchers for a week's rooms and meals, and a lightweight bicycle with saddlebags for your luggage. Cost of the tour is £30 inclusive and there are reductions for children.

THE ISLES OF EIRE

If you want to make a home in the isles off Eire, you must either be a citizen of Eire, or, through residence, take steps to become one. There may be exceptions, but as a general rule this applies. In this important respect residence on these islands differs from most other countries. Practically all of them belong to the Government of Eire, which periodically grants long leases to island-fanciers. The annual rentals in most cases are small—some off the coast of Mayo have been leased at as little as £7 ($17) a year—but almost all leases are granted either to citizens of Eire or to permanent residents. It is necessary to live in Eire for some time in order to qualify for a lease, but the time varies, depending no doubt on the official dealing with the inquiry and on the applicant.

Exceptions are sometimes made for British non-residents by granting them shooting and fishing rights on a particular island, renewable annually. This exemplifies the love-hate relationship the Irish have for the English—collectively they hate the English (or at least a large percentage of them do), but individually they often shower affection on English people and do them favours they would not do for persons of any other race. Of course, it helps greatly if you are a Catholic. It must be admitted that shooting and fishing rights give one practically all the privileges of island ownership, but there is always the insecurity of a year-by-year tenancy to contend with. On the other hand the comparative cheapness of the rentals makes these islands ideal for holiday homes. All applications should be made to the Government of Eire.

The rentals are 'comparatively cheap' in relation to most other parts of the world. You are not likely to come by a £7 ($17) a year lease very often and, if you do, you can bet on it that your island will be little more than a windswept rock or sandbank. Any worthwhile island will cost considerably more and, if you wish to buy from one of the relatively few private owners, you must be prepared to count in thousands rather than hundreds.

Nevertheless, Eire offers islands for sale at reasonable prices. For example, not so long ago Sandy Cove Island, in a sheltered inlet 3 miles from Kinsale, County Cork, was up for sale at £5,000 ($12,000). A 13 acre island, with its own boat slip, no house or building was included in the price. It had one advantage: at low tide you can walk to the mainland, though 3 miles at low tide seems quite a stretch.

The west coast of Eire is prolific in islands and the pleasantest are situated in the south-west. The climate here is mild, but the rainfall heavy. All the islands are tax-free. Many are uninhabited, but a fair percentage have cottages and even large houses to let or for sale.

To learn about island life off the Irish coasts, visit the Blaskets, off the coast of Kerry. Great Blasket is the most westerly of Eire's islands. You anchor off its shores, sound a foghorn and shout 'Naomhog!' ('Bring a boat'.) The inhabitants officially left the island in 1954, but some of them return in summer, and recently an American, Mr Taylor Collings, has made his home there and interested himself in developing the Blaskets for tourists. Those people of the Blaskets who return to the islands are specially hospitable to strangers and, like the Arabs, they show their esteem by killing a sheep in honour of their guests, and even in so isolated a group of islands you can be sure of a full table with plenty of eggs, butter, milk and soda bread. But left to themselves they feed frugally. As in the Outer Hebrides the islanders believe in fairies and you will often find boys dressed in kilts to prevent them from being carried off by the 'little folk', legend having it that the fairies ignore little girls.

Finally—and in 1972 it may sound rash advice—there are islands in Northern Ireland which, strictly speaking, should be listed in Chapter 2 on British islands. Well, politics and gunmen apart, Ireland is indivisible and the fact is that in the remoter country districts of Northern Ireland there are few signs of civic strife. Ultimately peace must come to this unhappy area of Ireland and it may be that an island bought at a bargain price there today will be worth a lot more in another decade.

Jack Suckling was serving in the Army in Aden when he read an advertisement for the island of Lustybeg in Lough Erne, which was being offered at £4,500 (this was before the current troubles erupted). He bought it and took along his wife and their two daughters when he left the Army. His story is surely proof that island-owning requires character. He had to go to London on business for three months shortly after he took the island, and his wife and children were left alone there for that period—during which they were marooned by bad weather for ten days and became so short of food that they had to dig up the seed potatoes they had planted, the 400 gallon water-storage tank burst and flooded the house, the lough was frozen over, and in between rowing the children to and from school Mrs Suckling had to chop down trees and feed the fire.

They triumphed in the end, however, and built two holiday chalets on the island and started a guesthouse. By 1967 they had five small timber chalets and were earning a decent living.

THE FAR EAST

The Far East is perhaps not, politically speaking, the most salubrious of areas for island-hunting. I omitted it entirely from *Islands for Sale* for this very reason, just as I have still left out any islands under the control of totalitarian states. My definition of 'totalitarian' may seem somewhat arbitrary. I include Soviet Russia and her satellites, but give Yugoslavia the benefit of the doubt. Spain and Greece are equally given the benefit of the doubt because, by and large, it is possible to enjoy individual freedom in those territories if you keep clear of poliitcs. China, Haiti and some Latin American countries can also be regarded as totalitarian. I assume every island-lover is a democrat, or at least a benevolent autocrat.

Perhaps today Japan is one of the stablest countries in the Far East, and since the war has enjoyed a quarter of a century of democratic rule. Well, there are certainly many islands around the Japanese coasts and in the past few years several of them in Japan's Inland Sea between Honshu and the Island of Shikoku have come on the market. Their average price is around £8,000 apiece, but in some cases the price has been as high as £12,000 for a very modest combination of sand and rock, mainly because a number of these islands are reputed to be the site of buried pirate treasure. In most cases the islands are the property of Japanese families who lost their fortunes in the war.

The islands for sale are in the south-eastern waters of the Inland Sea, off the wild coast of Shikoku. A typical island, already purchased for £5,000 near Takamatsu, covers only about 10 acres, and has a fisherman's cottage capable of conversion into a comfortable home. There are two safe beaches, a miniature forest of palms and pine-trees, a natural water supply and space for vegetable-growing, pigs and poultry. Curiously, foreigners have shown more interest than the Japanese in buying these islands, and they have not been discouraged.

The conventional picture of Japan today is of vast cities packed with teeming millions bustling backwards and forwards while they fulfil Japan's economic miracle. But there is a rural Japan, enhanced by some of the greatest scenic beauties in the world, where life continues at a slower pace. This is the Japan that Lafcadio Hearn, that superb interpreter of Nipponese life, portrayed so vividly in his prose-pictures and short stories about the country. It is not entirely an outdated picture, especially in the country districts he describes so well. Before considering buying an island in Japanese waters it would be advisable to read his works.

NORTH BORNEO

When *Islands for Sale* was published, I received the following letter from Mr Harold C. Johnson, of Labuan, North Borneo:

> I have just read, re-read and am re-reading *Islands for Sale* and feel I must let you know what a godsend it is to me at the present moment. I am a solicitor by profession, aged seventy years this year, and am retiring.
>
> I own an island off Labuan in Brunei Bay, of which I enclose a photo taken from the Singapore plane [see p70]. I first lived there shortly before the Japanese occupation when I was turned out and subsequently interned by the Japs in Kuching for three and a half years. On the re-occupation I returned to the island, cleared the jungle and built another house and re-made flower, vegetable and fruit gardens. The island itself is a small coconut plantation and therefore should not be hard to sell. Now, owing to local economic conditions, I am unable any longer to get labour to look after the place. I require three at least and after a period of difficulty it has become impossible to get them. I have therefore vacated the island to my great sorrow.
>
> All my friends and visitors say I will never get another island

> anything like it. The swimming is perfect. I was thinking it was hopeless, but nevertheless planning a journey in search of one [another island]. I decided to wait for your book which I saw advertised and now I am able to make plans with a new hope. I cannot do with an expensive island, but can provide an income up to £5,400, which should not be necessary for me.
>
> I have all the proper complexes—love of solitude, swimming (a necessity), gardening, farming and a library which by good luck survived the Japanese...

Mr Johnson went on to talk of his quest for another island. He was hoping to make an exchange of islands, while doubting if this was possible. But he proposed at the age of seventy to make what amounted to a world tour in quest of islands—to the Greek isles, the Caribbean and the Great Barrier Reef. 'I have always been an escapist and the older I get the more confirmed I become'.

We exchanged letters over the next year or two and I tried to encourage Mr Johnson to tell the story of the island paradise he had created and how he had set about it, because no one I knew has even made an island—at least so small an island—so self-sufficient. His planning of food production on the island was masterly and it was a tragedy that the war destroyed his hopes. Then there was silence: I heard he had moved on to Australia and then to Fiji, but was never able to trace him again.

In recent years I have tried again, without success, to find out what happened to Mr Johnson. However, I have learned something more about his island. It is Pulau, 3 miles east-south-east of Labuan, 1 mile long and $\frac{1}{4}$ mile wide, and now belongs to a Labuan Chinese family who use it for weekends and picnics. The only buildings there today are a handful of thatched huts, with two resident caretakers. Apparently Mr Johnson went to the Far East some years before World War II, intending to live among the Burmese Karens, but, passing through Singapore, he heard of the island and bought it. He was an itinerant lawyer in North Borneo, picking up cases as he travelled and then returning to the island to work on them. Living in a thatched, native-style house, with a makeshift office and a magnificent library of books, Harold Johnson declined absolutely to comply with the rigid codes of living imposed on Europeans in the pre-World War II Far East. While his efficiency revealed itself in the manner in which he made his island almost entirely self-supporting, he maintained the habits of a simple life. He declined to adopt conventional attire for Europeans, wearing nothing but shorts or a sarong while on the island and walking barefoot.

THE FRENCH PENAL COLONIES IN GUIANA

I never expected to include the former French penal settlements (of which Devil's Island is the most notorious) in a guidebook for escapists, but time has proved that they now deserve some consideration. They form a small archipelago of three—Devil's Island, St Joseph and Royale Island—situated near the equator and close to the coast of French Guiana in South America. Here French criminals and political prisoners were sent to rot and to die. Alfred Dreyfus, the French Army officer convicted of treason and later proved innocent, was imprisoned here in 1895, and Henri Charrière, nicknamed Papillon, was one of the few who escaped, living to write a bestseller on his experiences.

The cells where prisoners were once chained to the walls in almost total darkness are now overgrown with creepers, and the jungle, with the assistance of wood-consuming insects, is slowly destroying all traces of the penal settlement on Devil's Island and the others, and their natural tropical beauty is returning.

Royale Island is only 8 miles from the mainland, and has six inhabitants, one of whom is Mrs Ian Thompson, formerly a draughtswoman employed by the Rolls-Royce company in Sheffield, who lives in a decayed old house. She first saw the island and fell in love with it when her husband was working at Kourou on the South American mainland, where the French have established a station for launching space rockets. She decided to settle on Royale Island, where, she says, 'I pay no rent, no rates and no income tax'. Her income comes from her painting, both in oils and watercolours, and as she is a skilled fisherwoman she saves money by catching red mullet, grouper and other fish for her table.

Royale Island is exotically beautiful, providing exquisite settings for her paintings, with hibiscus blooms in abundance, palm trees and bougainvillea. There is plenty of fresh fruit to pick—mangoes, bananas, apricots, limes and breadfruit. Bathing, however, is almost impossible because of the sharks, which were always the main defence against prisoners escaping in the old days. The cost of living is low—Mrs Thompson reckons it to be about £5.60 ($13) per week—but there are other hazards apart from the sharks in the form of tarantula spiders, snakes and wild pigs, though these do not worry Mrs Thompson and her two dogs. There is one restaurant on the island, mainly catering for visitors.

THE EASTERN ATLANTIC

The Canary Islands have in recent years undergone considerable development, though less than Majorca, Ibiza and some of the Balearics. Avoid the main islands of Gran Canaria and Tenerife and concentrate instead on Lanzarote, though prices have steadily increased even there, and £15,000 is not an uncommon price for a bungalow close to the beach. A beach site plot of land can cost as much as £3,500 ($8,400), but the cost of living is low, less than in Spain. Older houses suitable for restoration can be bought for as little as £1,000, while on the lesser islands (Hierro, Gomera and La Graciosa) land can still be bought for as little as 20p (48c) per square metre.

The contrasts in the Canaries are much more marked than in the Balearics. In Tenerife and the large towns you can enjoy every kind of sophistication, but in Lanzarote you are in another world—a curious landscape of hump-backed craters, looking part lunar and part Arizona Desert. The villages of Lanzarote take one back to life 100 years ago, with camels drawing the ploughs, vines sheltered from the wind by drystone walls, and peasants whose way of life has barely changed for centuries. The island has a fire mountain, well worth visiting, grottoes and volcanic caves and some excellent village restaurants where you can enjoy a four-course lunch for 50p.

THE VERY POOR MAN'S ISLAND

I would like to insert a few words of encouragement for the last-ditch, desperate, impoverished escapist who cannot afford an island in the conventional sense of that word, though such a man will never let lack of cash defeat him. He is usually different from the average island home-hunter in that his kind of escapism is more abstract: he would rather have a solitary rock he could call his own, even if he could not live on it, than have no island at all. Well, there are more than 5,000 islands around the British Isles, the vast majority of them no more than rocks, so there must be millions of rocks going abegging throughout the world. Many of them could be bought for a few pounds, or leased for a peppercorn rent, as could some sandbank islands.

Most of the rock islands are uninhabitable and some are inaccessible. On a few you could erect a small hut and, if you are an ornithologist or a lover of solitude, such a rock could be a weekend hideout. There are only three pieces of advice one can give to rock-lovers: you must study your charts and maps and then go to the nearest village or town and find out for yourself whether you can buy, you must make sure it is possible to land on your rock, and you must check that transport is available.

The most romanticised rock island I know is Redonda, off Antigua. It is a gaunt tall block jutting out into the Caribbean and its single saleable commodity is guano, or bird manure, used for phosphate production. In 1880 a trader landed on Redonda, then unclaimed by anyone, and presented it as a birthday gift to his son, who later became M.P. Shiel, the Irish romantic novelist. Shiel appointed himself 'King' of Redonda and, when he died, left the island to his friend, the poet John Gawsworth, who declared: 'I have never seen the island, I cannot afford to travel out there. But I keep up the tradition of the Kingdom of Redonda and hold court once a year with some of my "dukes", who are all literary figures intent on perpetuating the name of Shiel'.

Periodically the romantically minded Gawsworth bestowed Redondan titles on various people. He made me editor designate of the *Redonda Herald*, a newspaper that only had one issue. Dirk Bogarde, the actor, was made a 'Duke' after he and Gawsworth had served together in India during World War II: his appointment stated that the 'dukedom' was 'for military collaboration at Bally, Bengal'. J. B. Priestley, Rebecca West, Stephen Potter and Naomi Jacobs were honoured 'for services to Shiel', while Diana Dors was made 'Duchess of Redonda' because of her 'services to beauty and charity', more especially because of the help she gave to badly burned airmen undergoing plastic surgery.

In June, 1958, Gawsworth advertised in *The Times:* 'Caribbean kingship with Royal Prerogatives, 100 guineas'. The 'kingship' of Redonda was not disputed by the British Colonial Office, though it included the privilege of creating nobility, which Gawsworth described in *Who's Who* as his hobby. He explained the advertisement as follows: 'I am only making this offer because I want to raise enough money to complete a life of the late king, the Irish writer, M.P. Shiel. He left his rights on the island to me absolutely. Now of course I cannot sell the actual island kingdom because it has been included in the Caribbean Federation'. There were many inquiries, but no firm bidders for the island under these conditions. In 1967 Gawsworth announced that he had abdicated as monarch and was appointing Arthur John Roberts as the next 'King'. Gawsworth died a few years later.

In 1964 Ernest Hemingway's brother Leicester announced that he had founded a 'new territory' and wanted six active young Britons to become citizens. He had to omit Americans from his offer because they were barred from dual nationality. Called the Republic of New Atlantis, this island chosen by Hemingway for his experiment must be the smallest in our list. It was only 6ft by 12ft and lay 7 miles off the Jamaican coast, 30 miles south-east of Montego Bay.

'It didn't exist until we made it,' said Hemingway. 'Every island in the world belongs to somebody today, so I picked a submerged bank fifty feet down and beyond the three-mile limit and I built it up with iron piping, stones, bamboo and steel cables.' By towing out disused concrete-hulled navy ships and filling them with rubble, Hemingway planned to create an island $\frac{1}{2}$ mile long and a $\frac{1}{4}$ mile wide. The serious purpose behind his caper was to establish a marine aquarium and protect Jamaican fishing rights.

Selecting a submerged sandbank 50ft below water is a doubtful basis for creating one's own island, but there are possibilities in developing a rock or building up a sandbank. Some flat rocks could be adapted for growing vegetation, if only sea-flowers and bushes. Sandbanks can be given artificial protection by using wooden structures and stones and creating barriers of marram grass. One sandbank off the Norfolk coast, less than 200sq ft and only 3ft above sea level, has been turned into an island 400ft long and 60ft wide, now planted with marram grass, a variety of mosses, sea arrow grass, horned poppy, sea convolvulus and sea asters.

Much more could be achieved in rescuing land from the sea in this way, at the same time creating a miniature island of some beauty as well as a naturalist's paradise. In some instances the would-be island creator could usefully link up with local naturalists' or ornithologists' societies. Mr E. D. B. Russell, of Roydon, Norfolk, wrote to the *Sunday Times* on

26 January 1964, making some practical suggestions on these lines:

> With the extension of the three-mile limit for fishing, might it not be worth while considering the reclamation of some of the sandbanks in the North Sea?
>
> With modern equipment, sheet piling, Mulberry Harbour-type units, and so on, it should not be too difficult to raise at least part of many of these sandbanks above high-water level and form islands of 'Upland'. Alternatively, sea walls could be constructed and the land reclaimed as 'marsh' below high-water level, or a combination of both.
>
> As to the advantages: from the Goodwins coal could be mined, and natural gas, if not actually oil, is considered likely to be found below the North Sea off East Anglia. In time considerable areas of land fit for agriculture would become available. Navigation is likely to be improved and the inshore fishing areas vastly increased. However, perhaps the greatest advantage would be the fillip it would give to the steel industry and to the East Coast generally where Mulberry Harbour units could be constructed on idle slipways.
>
> In addition, there is the important point of national security. The fishing rights alone could well attract other nations to construct islands if we do not forestall them.

Land reclamation is important for any small over-populated nation like Britain, and a survey suggests that many new islands could be created round our coasts, some of which could be developed by the private owner. There are, of course, legal problems to be considered when developing sandbanks within territorial waters, and these will vary from place to place, according to local bylaws and threats to navigation, but they are not insuperable. A number of river and estuary islands have been so developed and saved from disappearance by private owners building defences against flood waters. Much information on such experiments can be obtained from naturalists' societies, and, for a first-hand report on the practicability of such island-creation or development, ask such preservation experts as the Norfolk Coast Conservation Committee.

One of the nicest examples of do-it-yourself island-making was the work of Mr Jan Sniechowski, of Ipswich, who created an island measuring 55ft by 14ft and gave it to his daughter as a twentieth birthday present (see p160).

Mr Sniechowski made the island on a stretch of the River Gipping at Bosmere Mill, Needham Market, Suffolk. He had been engaged in restoring an old mill by the river, and, as this involved dredging his stretch of the river, the idea came to him to utilise what he dredged up as the base for an artificial island.

> I wanted to give my daughter something original for a birthday present. We claim that it is the smallest island in Great Britain and we have named it after the biggest river in my native Poland, the Vistula. My solicitors have drawn up all the necessary papers and transferred the land to my daughter. There is a bridge across to the island and, to add a bit of scenery, I have built a little windmill on it and planted a weeping willow tree. The island will also have a coat of arms which has now been designed.

ISLANDS FOR THE TIMID

Finally, a word to the timid escapist, the person whose courage (or commonsense) does not extend to adventurous activity. Remember the small island in a Devonshire river (see p26). This kind of island is probably your best bet. In the British Isles alone there are many such islands, especially in the Thames, and some of them have all the amenities of civilised life. One such is Ditton Island, connected with the Surrey bank of the river by a tollbridge. It was taken over by a small colony of island-lovers early this century, long before the bridge was erected. Its bungalows, with their neat lawns running right down to the water's edge, provide a splendid view of a most attractive stretch of the Thames. When you get tired of watching the boats go by, you can cross the bridge and pop into the Swan Hotel for refreshment. The island itself is not for sale, but property there comes on the market from time to time. There is even a Residents' Association.

MISCELLANEOUS ISLANDS

Higher up the Thames are other more secluded islands, some of which can be bought or leased. You will even see the occasional City gent punting himself across to catch the train of a morning, and his wife may be waiting to ferry him back each night. Some time ago Pats Croft Eyot Island near Wraysbury was up for sale at a price of between £10,000 ($24,000) and £15,000 ($36,000)—much less than the cost of a modest West End flat—and nearby Home Island, a 1½ acre freehold, was sold a few years ago for £15,000. Home Island has a fixed land link with nesting swans as a deterrent to strangers.

This book is but the aperitif before the meal. If it helps to titillate your palate, it will not have failed in its purpose. The meal itself—finding one's own island—must inevitably follow.

APPENDIX 1

REFERENCE BOOKS FOR ISLAND HOME-HUNTING

BAARSLAG, KARL. *Islands of Adventure* (Robert Hale). For information on the various outlying islands the author has visited

BANFIELD, JAMES. *Confessions of a Beachcomber*

BRADFORD, ERNLE. *The Greek Islands* (Harper & Row, New York)

CONOVER, DAVID. *Once Upon an Island* (Crown Publishers, New York)

DANIELSSON, BENGT. *The Happy Island* (George Allen & Unwin)

FORBES, ROSITA. *Islands in the Sun*. About the West Indies

FROMAN, ROBERT WINSLOW. *One Million Islands for Sale* (Duell, Sloan & Pearce, New York and Little, Brown & Co, Boston)

IREMONGER, LUCILLE. *It's A Bigger Life* (Hutchinson). For a description of life on Pacific islands

LOCKLEY, R. M. *I know an Island* (Harrap). For information on islands around Wales, Scotland, and Heligoland

MOSELEY, MARY. *The Bahamas Handbook (Nassau Guardian)*

Pacific Islands Year Book (Pacific Publications Pty, Ltd, Union House, 247 George St, Sydney, Australia). The publishers also produce *Pacific Islands Monthly*

ROGERS, STANLEY. *Enchanted Isles* (Harrap). For information on the historical and literary associations of several islands

SAATTUCK, GEORGE BURBANK. *The Bahamas Islands* (Macmillan Co, New York)

APPENDIX 11

WHERE TO OBTAIN FURTHER INFORMATION

Chapter 2 The British Isles

Isle of Man Tourist Board, The, Douglas, Isle of Man

Scillonian, The, Scillonian Press, St Mary's, Isles of Scilly. News magazine of the Scilly Isles

Tourist Information Bureau, Jersey, Channel Islands

The leading London estate agents are most likely to have information on islands for sale or to rent, but particularly recommended are the following:

Harrods Estate Offices, 32 Hans Crescent, London, SW1

Knight, Frank & Rutley, 19 Hanover Square, London, W1

Ralph Pay, Lord & Ransom, 127 Mount St, London, W1

Estate agents throughout the British Isles deal with islands from time to time and the following may be recommended:

Cagney & Sons, James, 19 Maylor St, Cork, Eire

Domaille, R., Diamond Vinery, Vieille Rue, Baubigny, St Sampson's, Guernsey, CI

Field's Estate Agency, Grenville House, Grenville St, St Helier, Jersey, CI

Lovell & Partners, St Peter Port, Guernsey, CI

Pritchard & Co, John, Cathedral Chambers, Bangor, Caernarvonshire, N. Wales

Shellbridge Ltd, Perseverance Buildings, Leeds Rd, Shipley, Yorks. For Channel Islands

Chapter 3 The Great Barrier Reef

Australian Lands Sales, Ltd, 6 Half Moon St, London, W1

HM Queensland Lands Department, Brisbane, Queensland, Australia

Chapter 4 The Pacific

Cromptons, Barristers & Solicitors, Thomson St, Suva, Fiji

Malley, Mrs Phyllis, Apt 706, Jane Arms, 430 Kaiolu St, Waikiki, Hawaii

Nordman, Oscar, estate agent, Papeete, Tahiti

Pacific Islands Monthly, Pacific Publications Pty, Ltd, 29 Alberta St, Sydney 2000, Australia

Senerosi Travel Agency, POB 1407, Suva, Fiji

Chapter 5 The West Indies and the Western Atlantic

Bahamas House, Mount St, London, W1

Bahamas Ministry of Tourism, The, 23 Old Bond St, London, W1

Bahamas Properties Ltd, 6 Half Moon St, London, W1

Bermuda Development Corporation, High Holborn, London, W1

Christie, Harold G., Bahamas Real Estate, 309 Bay St, Nassau, Bahamas

Hampton & Sons, 6 Arlington St, London, SW1

Previews International Ltd, 10 Buckingham Place, London, SW1

West India Committee, The, 18 Grosvenor St, London, W1

Chapter 6 The Mediterranean

Euralliance Overseas Investments, Harleyford, Marlow, Bucks

Hampton & Sons, 6 Arlington St, London, SW1

Knight, Frank & Rutley, 19 Hanover Square, London, W1

Leaver & Co, Marcus, 36 Bruton St, London, W1

Melpond Estate Agency, Park Mansions Arcade, Knightsbridge, London, SW1

Previews International Ltd, 10 Buckingham Place, London, SW1

Tufnell & Partners (International) Ltd, 50 High St, Ascot, Berks

Chapter 7 The Greek Islands

Dem Pazarlis, 34 Panepistimiou St, Athens (153), Greece

Mantheakis, Alexis, Solonos 9, Kolonaki, Athens, Greece

Previews International Ltd, 10 Buckingham Place, London, SW1

Thorpe & Partners, Bernard, 1 Buckingham Palace Rd, London, SW 1W OQD

Chapter 8 The Indian Ocean

Mancham, James R., POB 131, Mahé, Seychelles

Chapter 9 North American Islands

CANADA Keith, Hervey, 181 Eglinton Avenue East, Toronto. Specialises in island properties in Ontario

Lorn Tudhope Agencies, 24 William St, Parry Sound, Ontario. Specialises in Georgian Bay properties

Ontario Journal & Tax Sales Register, Sovereign Publishing Co, 110 Church St, Toronto

Pemberton Realty Corporation, Ltd, 418 Howe St, Vancouver, BC

Superintendent of Lands, Department of Lands & Forests, Parliament Buildings, Victoria, BC

USA Bureau of Land Management of the USA, Department of the Interior, Washington, DC

Chamber of Commerce, Avalon, California. For enquiries regarding Santa Catalina Island

Chamber of Commerce, Key West, Florida. For Florida Keys islands

Conservation Department, State Office Building, Madison 2, Wisconsin. For state-owned islands in Wisconsin

Division of Information, National Park Service, Department of the Interior, Washington 25, DC. For information on Channel Islands, California

Griffing & Collins, Shelter Island, New York

Kusterer Brothers, Inc, 129 Church St, New Haven, Connecticut

Old Island Realty Service, 322 Simonton St, Key West, Florida

Previews Inc, 231 South LaSalle St, Chicago 4

Regional Forester, US Forest Service, Box 1631, Juneau, Alaska

Thaxter, Miss Rosamund, Kittery Point, Maine. For the Shoals group

Thousand Islands Bridge Authority, Collins Landing, Alexandria Bay, New York

Wells, Sinclair, Land Agent, State Department of Agriculture, Tallahassee, Florida

Chapter 10 Miscellaneous Islands

Chilcott, White & Co, 125 South End, Croydon, Surrey. For the Canary Islands

Melpond Estate Agency, Park Mansions Arcade, Knightsbridge, London, SW1. For the Canaries

Poynton Advisory Service, Apartado 195, La Laguna, Tenerife, Canary Islands

Rosa Marina Interatlas (Estates) Ltd, 39 South Audley St, London, W1. For Madeira

Swedish Travel Bureau, 7 Conduit St, London, W1

INDEX

Italic numerals indicate illustration pages